MINING IN SOUTH AUSTRALIA

A Pictorial History

Compiled by J. F. Drexel

Special Publication No. 3

Drexel, John, 1952—
Mining in South Australia.

Includes index.
ISBN 0 7243 6094 8.

1. Mines and mineral resources—South Australia—History. 2. Mines and mineral resources—South Australia—Pictorial works. I. South Australia. Dept. of Mines and Energy. II. Title. (Series: Special publication (South Australia. Dept. of Mines and Energy); 3).

622'.099423

ISSN 0726-1527

1882 · 1982
100

DEPARTMENT OF
MINES AND ENERGY

Keywords

Mining/Mines/History/Photography/South Australia.

Issued under the authority of

The Hon. E. R. Goldsworthy, M.P.
Minister of Mines and Energy

'Gold don't know rules,
Where she be, there she be;
Where she be, there bain't I.'

Oswald Pryor

CONTENTS

PREFACE

The establishment of the Department charged with responsibility for administration of mining and quarrying activity, and for fostering mineral resource development in South Australia, dates from the appointment of Henry Yorke Lyell Brown, FGS, JP, as Government Geologist on 1 December 1882. The appointment of H. Y. L. Brown and creation of the Geological Office was formulated under the jurisdiction of the Commissioner of Crown Lands and Immigration. This, in turn, developed into the Geological, Mines and Goldfields Department, the precursor of the Department of Mines, following proclamation of the Mining Act in 1894.

Virtually unaided and with limited financial support, Brown's achievements during his thirty year term are recognised as being quite remarkable. He compiled the first official geological map of South Australia, which was published in 1883, and of the Northern Territory in 1898. His 'Record of the Mines of South Australia', ten thousand copies of which were distributed at the Great International Exhibition of Mining and Metallurgy in London in 1890, is still a popular publication (as a fourth edition, 1908).

Dr L. Keith Ward, who succeeded Brown in 1912, recognised that 'the contribution of the mineral industry towards every part of the fabric of modern civilisation is intimate and essential'. Accordingly, geological surveys were conducted to provide knowledge of the State's mineral resources and to assist the progress of industrial growth and independence.

The expanded functions of the South Australian Department of Mines and Energy as constituted in 1982, and of the Geological Survey of South Australia with which it is intricately linked, have continued to evolve. While it is responsible for regulatory and administrative control of the mining and petroleum industries, the Department also assumes a positive role in promotion of exploration and development of mineral resources by the private sector.

In recognition of the Centenary of H. Y. L. Brown's appointment as Government Geologist, and the foundation of the Department of Mines and Energy, aspects of mining and processing in South Australia are presented in this volume through an assembly of photographs, many of which have not been published previously.

The volume will arouse the interest, not only of those of Cornish mining descent, but also of those concerned with minerals, mining and related aspects of South Australian and, indeed, Australian heritage.

R. K. Johns
Deputy Director-General
South Australian Department of Mines and Energy

INTRODUCTION

Glen Osmond was the cradle of Australian mining, and it was from Wheal Gawler that the first metal-bearing ore was shipped from Australia to England in 1841. But it was not until copper was discovered at Kapunda in 1842 that widespread interest in metal mining was aroused in South Australia. The discovery of copper in 1845 at Burra added financial stability to an almost bankrupt Colony and, from that time until 1860, it has been claimed that the whole of South Australia was directly or indirectly employed by the mine. Burra was then the world's largest producer of copper.

The Colony had taken ten years to achieve a population of twenty thousand, but it expanded to sixty thousand over the next four years. By 1848, minerals comprised two thirds of total exports, and shipping volume expanded eight times from 1844 to 1848.

Any doubts that may have existed as to the State being a major copper province were dispelled by further discoveries in 1860 at Moonta and Wallaroo. These two fields contained more copper than the total from all other South Australian deposits until the recent discovery at Olympic Dam on Roxby Downs Station.

Mining, thus, had a profound influence in the economic development of South Australia: on immigration, notably from Cornwall and Wales; on infrastructure, through growth of the City of Adelaide and of the towns serving the mining communities; in improved communications, ports and railways; in provision of capital for investment in other enterprises, as at Broken Hill; for the promotion of education, notably the University of Adelaide; and in development of social character.

Altogether, approximately one thousand named mines were opened and, duly, closed by the turn of the century; most have little more than their names to recommend them. Old-timers still recall wistfully, but knowingly, that many had closed 'because water beat 'em' and that much ore remains to be recovered—such is the perception that may be attributed to 'secondary mental enrichment'.

The wealth to be derived from copper could not last forever, however, and copper was overtaken by iron at the end of the First World War as the principal metal produced in South Australia.

The non-metallic and industrial minerals, construction materials, gold, uranium, opal, coal and natural gas, have also assumed considerable economic importance since that time. And, in 1982, there is a general expectation that major resource developments based on petroleum, copper, gold and uranium will provide new challenges and opportunity: the Cooper Basin and Olympic Dam hold prospects for mining and processing on a scale not previously experienced in South Australia.

SILVER, LEAD AND ZINC

Australia's first metal mines were developed on outcrops of silver-lead minerals in the foothills of the Mount Lofty Ranges at Glen Osmond in 1841.

Intense exploration in the ranges led to further discoveries of silver-lead lodes at Almanda (1850), Wheal Ellen (1857), Aclare (1859), Mount Malvern (1859), and Talisker (1862).

Several small deposits were later found in the Flinders Ranges, and culminated with the development of the Ediacara mineral field in 1869. Small silver-lead mines were also worked in the Olary region and on Eyre Peninsula.

The Moonta-Wallaroo Mines produced silver as a by-product of copper refining, particularly during the period 1890-1923 when over seventeen tonnes were extracted from local and some imported ore.

An estimated forty-seven tonnes of silver and sixteen thousand tonnes of lead have been recovered from ore mined in South Australia.

The Broken Hill Associated Smelters (BHAS) at Port Pirie, which treats lead concentrates from Broken Hill, has had a strong influence on industrial localisation in the upper Spencer Gulf region. Several small mines in the State have proved viable because the smelters were available to treat their ore.

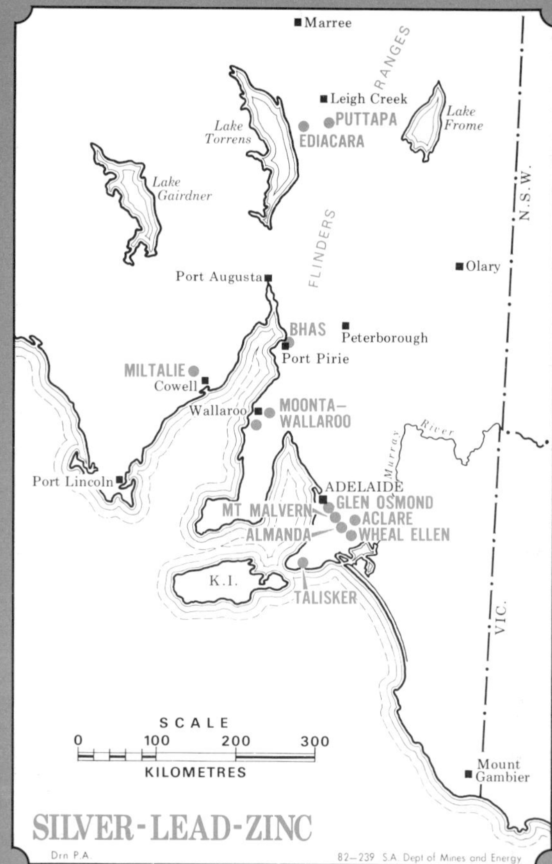

1 Woodley House, the former home of Colonial Treasurer Osmond Gilles, with Glen Osmond Mine to the right and Wheal Watkins to the left, circa 1900. The house was built in 1841 at about the time that Australia's first metal mine, the Wheal Gawler, was opened.

Glen Osmond

Galena (lead sulphide) was found on the property of Osmond Gilles, the Colonial Treasurer, in 1838, but the deposit was not the first to be mined. Two Cornish miners, Thomas and Hutchins, stumbled across a nearby outcrop of galena in 1840 at what became the Wheal Gawler, Australia's first metal mine.

In March 1841, the Wheal Gawler was taken over by Australia's first mining company, the South Australian Mining Association, but the company did not have the finance to develop the property, and operations virtually ceased within a few months. The property was sold to a German company in 1846, and subsequently to the Wheal Gawler Mines Association in 1848. They worked the mine until 1851 when most miners left to join the Victorian goldrush.

The galena discovery on Osmond Gilles' property was opened up in 1843 as the Glen Osmond Mine, known later as the Gilles Glen Osmond. Several tonnes of ore were raised and shipped to England for trial smelting. Apparently little further ore was mined until 1846 when the Glen Osmond Union Mining Co. was formed. The mine, when it closed in 1850, was the largest in the area.

The Wheal Watkins, situated between the Wheal Gawler and Glen Osmond Mine, was probably first worked in 1844, although some ore was earlier sent to England for trial smelting. The mine closed in 1851 as a result of the Victorian goldrush but was worked again in 1888-89, and the Tarcoola Development Syndicate did some prospecting in 1916.

Another group of silver-lead lodes was opened up in 1857 within sight of the old Tollgate on Mount Barker Road. These belonged to the Wheal Augusta, and were exploited by the Glen Osmond Union Mining Co. until 1850; a small amount of ore was extracted between 1854 and 1856. The Mount Osmond Silver-Lead Mining Co. reopened the workings from 1888 to 1889, but lack of capital forced the sale of property and plant in 1890.

The combined output of all mines in the district has been estimated at twenty-three hundred tonnes of ore, which yielded seventeen hundred tonnes of lead, and one and a half tonnes of silver.

2 A 1900 view of Wheal Watkins silver-lead mine, with concentrating machinery in the foreground.

3 *Wheal Augusta Mine and the Tollgate at Glen Osmond, circa 1900. The mine, one of the smallest in the foothills, produced only a few tonnes of ore. The white painted chimney of the Glen Osmond Smelter is at upper left.*

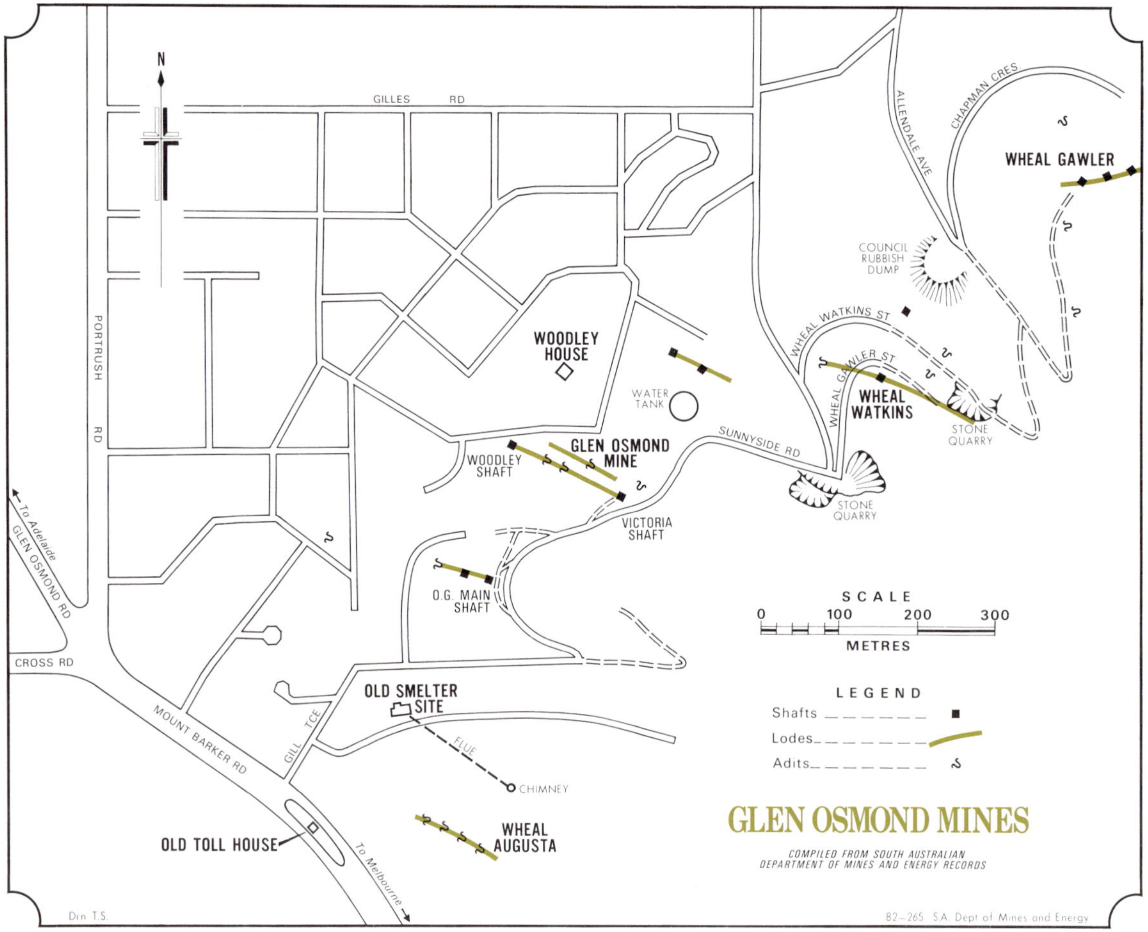

4 Adelaide in 1900, with ruins of the Glen Osmond Smelter in the right foreground. The smelter operated from 1849 to 1850, treating ore from the Glen Osmond mines and other mines in the Mount Lofty Ranges.

5 Part of the Glen Osmond Smelter being used in the 1950s as a farm shed.

Aclare

Silver and lead minerals were discovered near Kanmantoo at the site of the Aclare Mine in 1859, but mining did not begin until 1868. At first, tributers under the employ of F.C. Singleton worked on a small scale; in 1882, the Aclare Silver Mining Co. N.L. leased the mine from Singleton and spent a large amount of capital on development. Although the company raised rich carbonate ore, sulphide minerals which could not be processed were also encountered. The price of silver fell and the lease was surrendered in 1884. Several other attempts at treating the ore also failed.

By 1888, the success of the Broken Hill silver-lead mines was attracting English capital to Australian silver mining, and the Aclare lease, acquired by a syndicate of Adelaide investors, was offered for sale in London. The buyers, Kangarilla Proprietary Silver Mines of South Australia, raised sufficient ore to pay wages but had insufficient capital to develop the mine.

David Rosewarne, after resigning as Government Inspector of Mines in 1891, took over management of Aclare. He supervised underground development, erection of buildings, and installation of new plant and machinery. The mine had just begun to operate efficiently when severe drought dried up the water supply. As the ore processing was water dependent, the mine was forced to close.

A new ore-smelting furnace, designed to handle the sulphide ore, was erected in 1895, but the price of silver dropped again and Aclare was not worked after 1899. Total production since 1868 was approximately twelve hundred tonnes of lead and fifteen and a half tonnes of silver.

In 1965, the mine was acquired for the training of mining engineering students at the South Australian Institute of Technology.

6 Ruins of the Aclare silver-lead mine in 1981. The multi-storey engine house was built in 1891 and housed two Cornish boilers, an engine room, and a machinery room. The engine room had one engine to operate the crusher and rolling mill, and another to operate the jiggers and concentrator. Ore, trucked to a hopper on the roof, gravitated through the various processing stages. The furnace and chimney were built in 1895.

Talisker

An outcrop of silver, lead and arsenic minerals was discovered east of Cape Jervis by the McLeod family while searching for gold in 1862. The find was named Talisker after a town on the Isle of Skye. The Talisker Mining Co. was established in late 1862, with twenty men under the management of Captain Price.

Seven shafts were sunk along the main lode, the deepest to about one hundred and thirty metres; a poppet head was erected over the main shaft and horse whims over the others. By 1865, a flue, furnaces, engine and crusher house, and reservoir had been constructed; a ten-head stamp battery was added in 1867. A large kiln was built to make fire bricks from local clay.

Two mining towns built near the mine, Silverton and Silverton Extended, were quite substantial with shops, restaurant, hotel, post office, general store, day school, and night school for youths employed at the mine during the day. A Wesleyan Chapel, built at Silverton in 1866, held services until 1875.

Ore and ingots were carried several kilometres by either bullock or horse-drawn wagons to a small beach near Cape Jervis, named the Fishery, for shipment to England. A total of twenty-four thousand bags or ore, dressed and bagged on the beach, and eight hundred and ninety tonnes of silver-lead ingots, were manhandled through chest-deep water to lighters for ferrying to the larger ships.

Mining operations were suspended in 1871 due to lack of finance and excessive underground water.

The mine was re-opened in 1889 under Captain Tresize. A new adit, large enough to accommodate a double tramway, was driven for more than four hundred metres. Arsenic was recovered from the ore, and two tonnes of arsenic minerals were also extracted from the old smelter flue. Operations ceased in 1891.

The mine was later worked by small companies or individuals. When the mine closed in 1935, underground development totalled more than twelve hundred metres. Estimated total silver production from 1868 to 1935 was four and a half tonnes.

The Talisker Mine and environs were purchased in 1976 by the State Planning Authority as a recreation park.

7 Talisker silver-lead mine in 1889. The main shaft and engine house are in the centre, and Captain Price's cottage is on the hill behind. In the right foreground from left to right are the blast furnaces with twin feed hoppers, ore-sorting shed, crushing house and ten-head stamp battery. Campbell's Creek, with the sea in the distance, is at far left.

8 *A thatched-roof miner's cottage near Talisker Mine in 1869.*

9 A boiler on its way to Talisker Mine in the late 1800s.

10 Remains of the ten-head stamp battery in 1917. A mixture of ore and water was fed below the stamps which were raised and dropped by cams on the wooden axle. This crushed the rock to a powder suitable for ore-mineral concentrating. Driving power for the machine was provided by a steam engine in the building at left.

11 Inspecting the main shaft during 1924. The view is southerly towards Campbell's Creek.

Ediacara

The Ediacara mineral field, west of Beltana, comprises three principal mines—Beltana, Ediacara, and Black Eagle.

The first was the Beltana, found by a Mr Flett in 1869. Mining of the silver-lead ore, directed by Captain Terrell from Blinman Mine, continued until 1892, and again from 1901 to 1903. The mine was also known as Warioota, and Beltana Blocks.

In 1888, Ediacara Mine, also known as Ediacara Consols and New Ediacara, was opened up on an outcrop of silver-lead minerals discovered by W. B. Greenwood.

The small township of Ediacara, built on the field in 1891, provided services only until 1892 when both the Beltana and Ediacara mines closed.

Ediacara Mine was worked periodically between 1903 and 1913, and a smelter erected in 1907 produced lead bullion until 1909.

Black Eagle Mine, discovered by A. Gordon in 1907, was worked periodically for copper, and abandoned in 1918 after yielding fifty tonnes of copper metal.

Exploration in the 1960s and 1970s failed to locate further economic mineralisation.

A Department of Mines geologist discovered rare 650 million year old fossils, including jellyfish and worm imprints, near the Beltana Mine in 1946; the discovery site was declared a Fossil Reserve in 1958.

12 Ediacara lead smelter in 1909. Slag produced during smelting was dumped in front of the smelter building by large cast iron ladles, several of which are on top of the dump.

Wheal Ellen

Nearly one hundred men were employed at the Wheal Ellen, near Strathalbyn, for the first few years after mining commenced in 1857. A large amount of capital was spent on development and treatment works, but ore sales were insufficient to cover costs and the mine closed in 1866.

After reopening in 1906, silver, lead and gold ores were produced until 1908 when the Commonwealth Silver-Lead Co. Ltd of Sydney acquired the property and installed a stamp battery and concentrating table. Silver, lead and gold were extracted, and five thousand tonnes of pyrite were mined and sold to sulphuric acid manufacturers in Adelaide.

The mine closed in 1911 after producing an estimated five and a half tonnes of silver, twenty-one hundred tonnes of lead, and twelve and a half kilograms of gold.

13 The Wheal Ellen, circa 1908. Fine material from the steam-driven five-head stamp battery, in the right foreground, was passed over a concentrating table to extract gold and silver-lead minerals.

Miltalie

Several small lead and copper deposits were worked near Miltalie in the late nineteenth century, but none proved large, even by South Australian standards.

A good deal of excitement was created in 1912 when Mr Atkinson, a local prospector, revealed the location of a new silver-lead find. A small shaft was sunk, and several specimens assayed. The richest contained over sixty kilograms of silver per tonne, but extensive prospecting failed to locate a significant lode.

15 Another small silver discovery site near Miltalie in 1912.

14 Atkinson's 1912 silver-lead discovery near Miltalie.

Puttapa

Willemite (zinc silicate), in association with a variety of lead minerals, was discovered at Puttapa in 1966 during the course of regional exploration for base metals. Open cut mining has been undertaken by Electrolytic Zinc Co. (Australasia) Ltd, and the ore exported according to demand.

Similar, smaller, unexploited deposits have been discovered at Aroona and Third Plain.

16 Northerly view of Puttapa Mine in 1975, with the mine camp in the background.

The Broken Hill Associated Smelters

Port Pirie was established as the seaport for the massive silver-lead-zinc ore bodies discovered at Broken Hill in 1883. At first, ore was carted by bullock wagon, and later by rail, to the wharf facilities, the first load arriving at the docks in August 1885.

The British Broken Hill Pty Ltd built smelters at Port Pirie to treat Broken Hill ore concentrates in 1889. In the same year, The Broken Hill Proprietary Co. Ltd (BHP) constructed a lead refinery adjacent to the smelters to treat bullion produced on-site at its mine. BHP was the largest of the eight Broken Hill producers, and bought the British Smelters in 1892; the BHP smelting operations were subsequently transferred to Port Pirie. Local supplies of flux for the smelters were obtained from ironstone deposits at Iron Knob.

The ore mined at Broken Hill from 1885 was mainly lead carbonate and sulphate, suitable for direct smelting. However, with deepening of the mines at the turn of the century, lead and zinc sulphide ores became dominant. As these had to be roasted before the metals could be extracted by smelting, BHP installed a sintering plant at Port Pirie. The other Broken Hill producers, with no smelting or refining facilities, had to be satisfied with overseas sales of ore concentrates.

A zinc smelter was built by BHP in 1906 to treat the large amounts of sphalerite being received. Two years elapsed before the plant operated efficiently, but it was closed in 1921 as a non-economic venture.

The First World War had a great impact on the smelters. BHP, with dwindling ore reserves, had ceased to be the largest Broken Hill producer and was considering a changeover to iron ore smelting and steel manufacture elsewhere in Australia. The other mines lost their European markets because of trade embargoes, and a partial shutdown at Broken Hill looked likely. The solution was for a co-operative to take over the Port Pirie smelters and treat all Broken Hill ore. The Broken Hill Associated Smelters Pty Ltd (BHAS) was formulated with this capacity in 1915 from three shareholding supply companies including BHP.

17 Port Pirie smelters, circa 1903. Rail-cars filled with Broken Hill ore concentrates are in the foreground. Modern fume filtration systems have eliminated the pollution problems evident when this photograph was taken.

18 *Sketch of Port Pirie made from a hot-air balloon by J. J. Christie in 1896, with the lead smelter and refinery in the foreground.*

19 *A similar view in 1980. The two hundred and five metre chimney stack is the second tallest in Australia.*

The facilities were expanded to handle the extra ore concentrate input, and within a few years the company was the world's largest lead producer, and still operates the world's largest lead smelter and refinery. BHP, which sold its BHAS interest in 1925, continued to send concentrates to the smelters until its mine closed in 1939.

The BHAS research department has introduced many innovations to lead smelting, the best known being the de-silverising kettle which enables silver and gold to be extracted continuously from molten lead bullion.

In 1967, a plant was commissioned to extract and refine zinc metal from lead blast furnace slag and the six million tonnes of granulated slag accumulated from decades of smelting. Approximately half of the old slag dump had been reprocessed to the end of 1981.

The opening of the standard gauge railway link between Broken Hill and Port Pirie in 1970 simplified shipment of ore concentrate; up to ninety per cent of BHAS feedstock is carried on this line.

Annual smelter production capacity is now two hundred and fifty thousand tonnes of lead, two hundred and eighty tonnes of silver, and forty-five thousand tonnes of zinc. Nearly seventy-five thousand tonnes of sulphuric acid are recovered from sinter gases each year for sale to fertilizer manufacturers. Cadmium, copper matte, gold, and lead alloys are important by-products.

20 The lead blast furnace feed floor in 1904. Roasted ore concentrate, flux, and coke were shovelled into the furnaces below. This system was replaced by telphers in 1929.

21 The telpher blast furnace loading system in 1936. The telphers eliminated the need for smeltermen to charge the furnaces manually; hoods were placed over the furnaces to prevent release of toxic fumes, and the number of reported lead poisoning cases among smeltermen dropped dramatically.

22 Lead blast furnace in 1908. On the left, lead bullion ingots are being cast directly from the furnace; a worker is pouring slag into a ladle on the right. The pipes above the lead moulds inject air into the furnace to enhance liberation of lead from the charge.

21

23 *Blast furnace building in 1915. The horse-drawn ladles are carrying molten slag.*

24 *Southerly view of Port Pirie smelter's waterfront, circa 1900. Lead stacks are to the right of the railway wagons; in the foreground are cones of moulded slag which may have been used for the construction of a breakwater or seawall.*

25 *Thirty-six thousand tonnes of lead ingots awaiting shipment in 1918.*

26 These electrically driven Newnam trucks, photographed in 1925, were the fore-runners of forklift trucks.

27 The silver weighroom, circa 1910. Each fifty pound silver bar was stamped with an identification number, weighed, and recorded by the clerk on the right.

28 This disastrous fire in 1921 came only one week after an eighteen month strike by miners at Broken Hill. Low ore concentrate reserves and loss of the smelters seriously affected lead production for several years.

29 An easterly view of the BHAS in 1959. The large black dump in the centre is lead blast furnace slag which has been reprocessed for zinc since 1967. The Port Pirie uranium treatment works, which recovered uranium oxide from Radium Hill ore, is in the foreground.

30 Southerly view of Port Pirie waterfront in 1892. Lead ingots are stacked next to the ship.

31 Modern wharf facilities in 1980. Lead ingots are stacked in the foreground, with stockpiles of zinc concentrate beyond. The tippler system in the right foreground, which handles incoming rail traffic, automatically weighs and empties each rail wagon, then samples the lead and zinc concentrates for assaying.

COPPER

The early history of economic development in South Australia was dominated by the discovery and development of copper deposits, and by the 1860s the State had earned the title 'Copper Kingdom' by virtue of some of the largest copper mines in the world.

The first major discovery was at Kapunda in 1842, several small lodes were found in 1843 at Montacute, and a large outcrop of copper carbonates was found at Burra in 1845.

These finds stimulated exploration, and the Mount Lofty Ranges were prospected intensively, resulting in the 1845 discoveries of copper-gold at Kitticoola and copper at Kanmantoo.

The next major finds were at Moonta in 1860, and Kadina (the Wallaroo Mines) in 1861. These were to rank amongst the largest copper producers in the world; production was continuous until 1923, and their combined output of over three hundred and thirty four thousand tonnes of copper metal is greater than the total output from all other copper mines in the State.

Blinman Mine, discovered in 1862, was to become the sixth largest copper producer. Other significant deposits were found at Yudnamutana (1862), Prince Alfred (1866), Sliding Rock (1870), Mount Gunson (1875), Clara St Dora (1880), Mutooroo (1890), and Lorna Doone (1892).

Re-examination of old mining districts in the 1960s, at a time of demand for copper, led to the reopening of several old mines, and to new discoveries of ore. Burra commenced production in 1969, Kanmantoo in 1970, and the Cattlegrid orebody was developed near Mount Gunson in 1974.

By far the most significant South Australian copper discovery made this century is that at Olympic Dam on Roxby Downs Station near Andamooka. Drilling in 1975 revealed a large body of copper-gold-uranium mineralisation, and feasibility studies are proceeding.

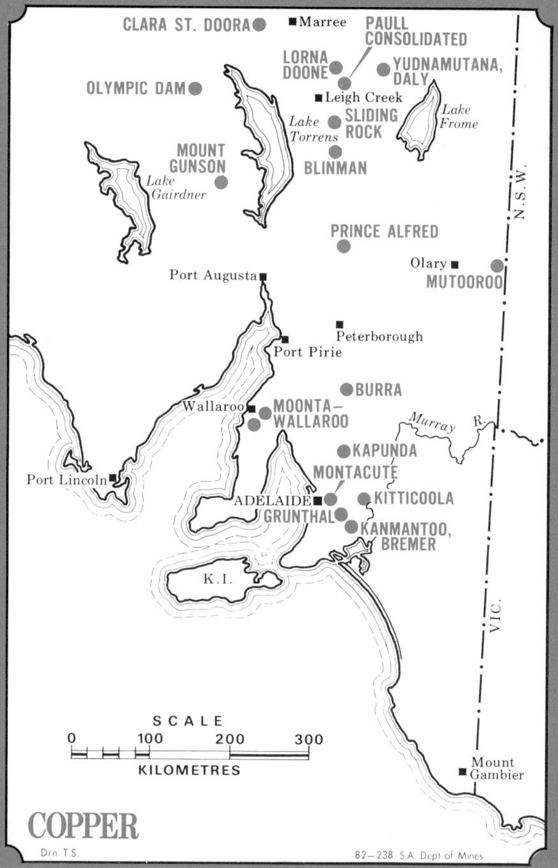

Kapunda

In 1842, Francis Dutton discovered green copper ore while searching for stray sheep. He discussed the find with his neighbour, Captain Charles Bagot, who was already aware of the lode. As neither man owned the land, Dutton and Bagot purchased a section around the outcrop at the Government price of £1 per acre.

The first ore was raised in 1844 at Australia's first major copper mine, Kapunda, by simply shovelling the outcrop of blue and green copper minerals into carts; the ore needed little further dressing. By December there were a dozen Cornish miners at work, underground mining had begun, and a small mining town was developing.

The underground workings reached the water table in 1845, and a horse whim was installed for bailing the water which was used in a crude ore dressing process. By 1847 the whim was incapable of handling the volume of water encountered in deeper workings, and a steam engine was purchased in England.

For the first five years, ore was carted to Port Adelaide on drays for shipment to smelters at Swansea in Wales. In 1849, the first smelting furnace at Kapunda was fired up, followed shortly after by a second. A refinery, added to the smelters in 1861, enabled a high grade copper product to be manufactured on site.

By 1851 Kapunda was a prosperous mining town of more than two thousand people, but the Victorian gold rush severely affected its population and the mine closed until 1855. After this, the mine developed rapidly, reaching peak production in 1857.

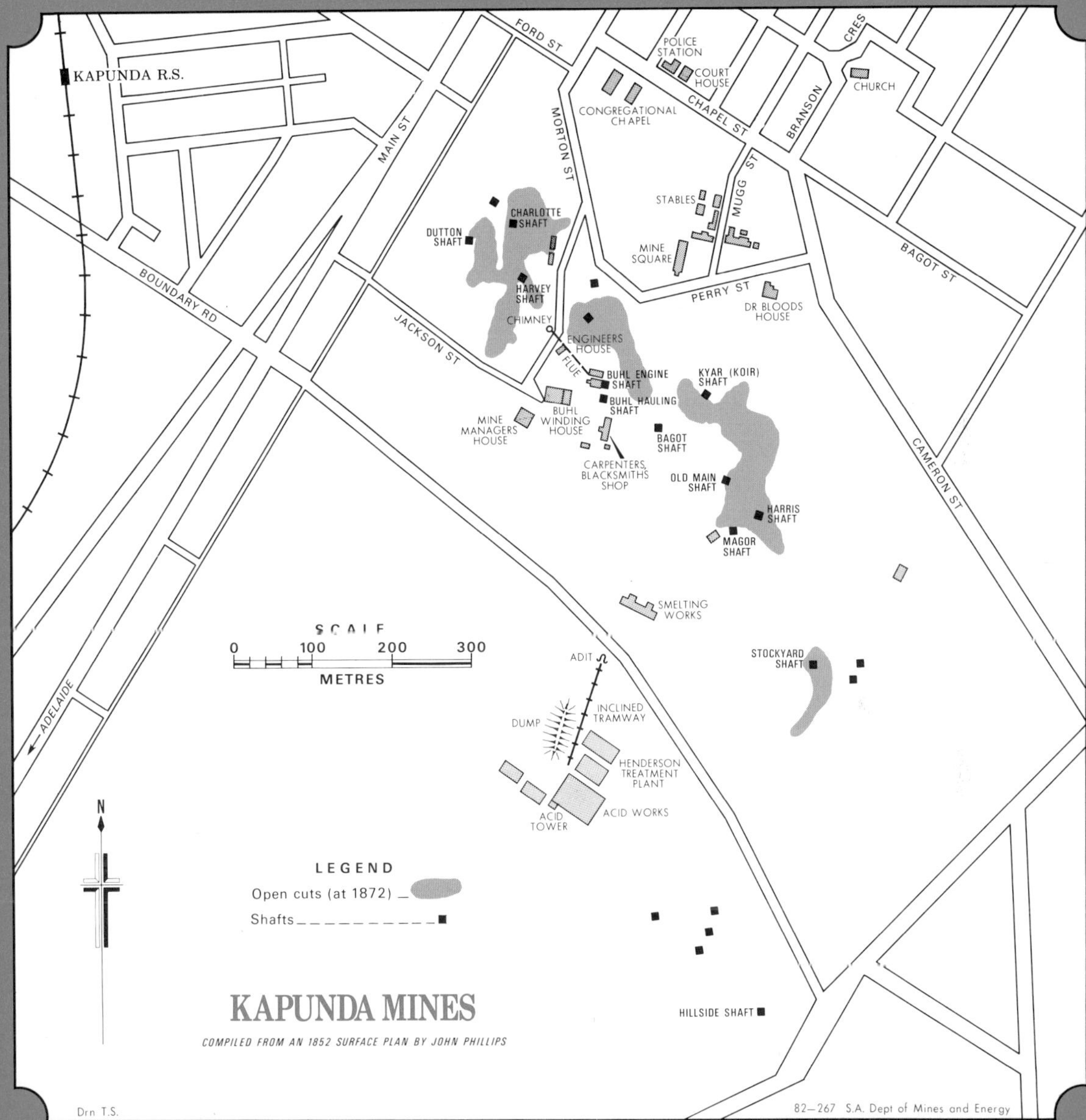

Kapunda was a tributers' mine: the Kapunda Mining Co. took a royalty on the value of ore raised, and exercised practically no supervision over the operations. The result was a series of burrows, often less than a half metre in width, without system or method.

After 1863, with exhaustion of the richest lodes, the higher cost in raising less accessible ore, and a fall in the copper price, financial losses were heavy. A Henderson treatment plant was erected in 1867 to concentrate and dissolve low grade carbonate ore in acid, and precipitate copper on scrap iron. Ore was mined from open cut and underground, and loaded into trucks which were hauled up an inclined tramway to the treatment plant. The process was an economic failure, however, and the mine closed in 1879.

Kapunda was worked periodically until 1912 by tributers who gouged small parcels of ore above the water table.

Total production from 1844 approximated fourteen thousand five hundred tonnes of copper metal.

32 Kapunda copper mine, circa 1875. The Henderson ore treatment plant is at left, mine workings in the centre, and miners' cottages at right. The house of the mine doctor and first mayor of Kapunda, appropriately named Dr Blood, is in the right foreground. Parts of the stone dwellings of Mine Square, at far right, remain today as Australia's oldest mine buildings.

33 Kapunda Mine in 1875. In the foreground are the crushing, leaching and precipitation sheds erected in 1867 as part of the ore treatment plant. The smelters with their square chimneys are in the right foreground, and behind is part of the main open cut. Kapunda township, with its prominent church spire, stands on the hill.

34 Smelters (left) and the ore treatment plant pictured during the same year. The wide tower at the ore treatment plant was used to condense sulphurous fumes released during burning of pyrite, one of the gangue minerals. Sulphuric acid was distilled from the resultant solution and used to manufacture hydrochloric acid for leaching of low grade copper ore; copper was precipitated as a black powder on scrap iron. Leaching and precipitation were carried out in the two sheds in front of the treatment plant. Part of the inclined tramway from the forty fathom underground level is to the right of the smelters in front of the leaching sheds. Ore trucks were hauled up the tramway by a stationary engine.

35 Derelict mine workings at Kapunda in 1924. The building at right is a winding house; in the right middle distance is the rectangular acid condensing tower of the old ore treatment plant.

Burra

Originally known as Monster Mine, Burra was discovered in 1845 by Thomas Pickett, a shepherd; the find came shortly after William Streair's discovery of copper minerals a few kilometres south at what was to become the Princess Royal Mine. To obtain mineral rights to both deposits, land had to be purchased from the government for £20 000. This large amount was raised by two parties; one was a group of capitalists and pastoralists nick-named 'The Nobs', and the other comprised tradesmen from Rundle and Hindley Streets in Adelaide known as 'The Snobs'. After purchase, lots were drawn to determine ownership of the two mines. The Nobs drew Princess Royal, an unfortunate lot to win because mineable tonnages of ore proved to be low, and the mine closed in 1851.

The Snobs, otherwise known as the South Australian Mining Association (SAMA), won the Monster Mine. Underground mining began in September 1845 and, about a month later, the mine was renamed Burra Burra (often shortened to Burra) after the nearby creek. Hand-picked ore was carted by bullock wagon to Port Adelaide, and from there shipped to Swansea in Wales for refining to copper metal. Later, ore was carted to the nearest railhead at Kapunda for transport to Port Adelaide.

Captain Roberts was appointed as first mine captain and held the position until 1847 when Captain Henry Roach was appointed mine superintendent. Captain Roach was supported by 'grass' and 'mine' captains. Grass captains managed the ore-sorting yards and surface workings, and the mine captains ensured that the mine was safe and ore was being extracted economically.

Large smelters were built near Burra Burra Creek in 1849 to save transport costs. The Smelts, as they were known, consumed over six hundred tonnes of firewood weekly, which almost totally denuded the landscape around Burra of trees. Later, Newcastle coal was brought in through Port Wakefield to supplement local firewood and, in 1851, the Smelts employed one thousand labourers and four thousand bullocks for cartage.

Water was always a problem in the mine as the water table was only thirty-six metres below the surface, and buckets attached to horse whims made little impression on the water level. A pumping engine was installed on Roach Shaft in 1849 for dewatering to enable deeper stopes to be worked. An even more powerful engine, installed on Schneider Shaft in 1852, was not commissioned immediately because the Victorian gold rush stripped Burra of almost the entire workforce. The Roach Shaft pump was stopped to reduce mining costs, and ore was extracted from upper levels only.

Miners began returning to Burra in late 1854, and the new pumping engine on Schneider Shaft was started a year later. Even this could not dewater the entire mine, and another pump was installed at Morphett Shaft. The large volume of water now raised was used to drive several water-wheel-powered crushers.

In 1867, John Darlington, a mining engineer from England, recommended open cutting rather than underground mining to enable lower grade ore to be treated, and high grade ore pillars to be recovered. SAMA approved his recommendations.

The mine surface was cleared of horse whims and buildings in 1869, an ore-dressing tower and concentrating plant were erected, and an incline railway constructed to the working floor. The Smelts were almost completely disassembled, and most of the furnaces and equipment shipped to new smelters being constructed at Newcastle, NSW. From 1870, most of the Burra concentrates were smelted at Port Adelaide.

Open cut operations began in 1870 and continued until 1875 when SAMA realised that open cutting alone would soon not be viable. Underground operations recommenced below the old levels and were successful for a time, but high dewatering costs and depressed copper prices forced the mine to close in 1877.

Apart from a small amount of tributing, the mine remained dormant until 1961 when the Department of Mines began investigations. These were followed up by Mines Exploration Pty Ltd and Samin Ltd, with the latter recommencing open cut mining in 1971. The mine was purchased by Adelaide & Wallaroo Fertilizers Ltd in 1978, and worked until exhaustion of ore in early 1981.

Dewatering was the biggest problem at Burra. Between 1845 and 1877 more than one and a half thousand tonnes of water were removed for each tonne of copper smelted, while from 1976 to 1981, four hundred and thirty tonnes were removed for each tonne of copper metal equivalent. Since 1845, nearly one hundred billion litres have been pumped from the mine.

Approximately seven hundred thousand tonnes of ore containing fifty thousand tonnes of copper were raised to 1877, and two million tonnes of ore containing forty thousand tonnes of copper were mined during 1969-81.

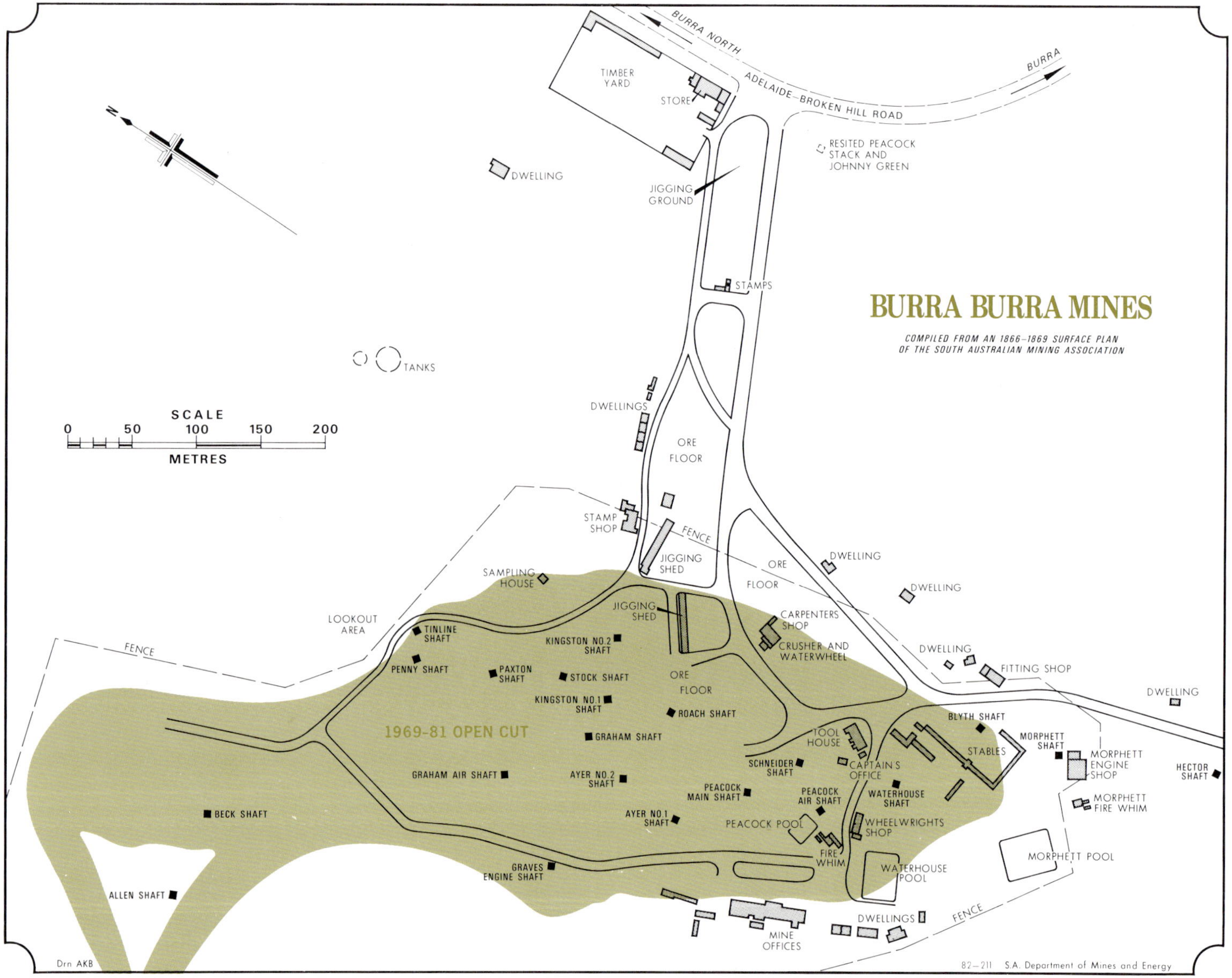
BURRA BURRA MINES
COMPILED FROM AN 1866–1869 SURFACE PLAN OF THE SOUTH AUSTRALIAN MINING ASSOCIATION
BURRA NORTH
ADELAIDE–BROKEN HILL ROAD
BURRA
TIMBER YARD
STORE
RESITED PEACOCK STACK AND JOHNNY GREEN
DWELLING
JIGGING GROUND
STAMPS
TANKS
SCALE
0 50 100 150 200
METRES
DWELLINGS
ORE FLOOR
STAMP SHOP
FENCE
JIGGING SHED
ORE FLOOR
DWELLING
DWELLING
SAMPLING HOUSE
LOOKOUT AREA
JIGGING SHED
CARPENTERS SHOP
CRUSHER AND WATERWHEEL
DWELLING
FITTING SHOP
DWELLING
TINLINE SHAFT
KINGSTON NO.2 SHAFT
FENCE
PENNY SHAFT
PAXTON SHAFT
STOCK SHAFT
ORE FLOOR
KINGSTON NO.1 SHAFT
ROACH SHAFT
1969-81 OPEN CUT
GRAHAM SHAFT
TOOL HOUSE
BLYTH SHAFT
MORPHETT SHAFT
STABLES
MORPHETT ENGINE SHOP
HECTOR SHAFT
SCHNEIDER SHAFT
CAPTAIN'S OFFICE
GRAHAM AIR SHAFT
AYER NO.2 SHAFT
PEACOCK MAIN SHAFT
PEACOCK AIR SHAFT
WATERHOUSE SHAFT
MORPHETT FIRE WHIM
BECK SHAFT
AYER NO.1 SHAFT
PEACOCK POOL
WHEELWRIGHTS SHOP
FIRE WHIM
MORPHETT POOL
WATERHOUSE POOL
GRAVES ENGINE SHAFT
ALLEN SHAFT
DWELLINGS
FENCE
MINE OFFICES
Drn AKB
82–211 S.A. Department of Mines and Energy

36 The Smelts, with Burra Mine in the distance, painted by S. T. Gill in 1850.

37 Southerly view of Burra copper mine in 1869. In the centre background is Peacock engine house, with lines of wooden frames carrying flatrods to operate dewatering pumps in the mine shafts. Morphett engine house and winding house are in the left background behind a launder carrying water pumped from a shaft to the water wheel at the crushers. Shingle-roofed ore sorting sheds are in the left foreground. Horse whims used for raising ore are in the right foreground, and behind these are the extensive mine offices. The mine was being prepared for the change-over from underground to open cut operation at this time. Ropes have been taken off the horse whims, and some of the flatrods leading from Peacock engine house have been removed from their frames.

38 A similar view in 1877. The open cut is to the right in front of the mine offices. Ore was hauled up the incline in the centre background, and waste material dumped on the mullock heap behind. The launder and waterwheel which drove the ore crushing plant are left of the incline; in front are the now derelict ore sorting sheds.

39 Burra open cut in 1877. Ore and mullock was hauled from the open cut in trucks which ran on tracked inclines; two trucks are on the main incline in the foreground. Haulage power was provided by Peacock steam engine. The mine closed a few months after this photograph was taken.

40 Part of the open cut in 1877. Many of the old timbers used underground between 1845 and 1868 have been exposed. The pit is almost worked out, and remnants of ore are being extracted by underground methods. An ore truck is emerging from a drive behind the miner standing on the railway track at lower left.

41 Ruins of The Smelts, circa 1899.

***42** Morphett engine house, winding house, and launder in 1892. Morphett pump, started up in 1860, was capable of lifting sixteen million litres of water daily. This was run along launders to waterwheels at the crushers and in the pit. Excess water was stored in a reservoir in the right background. The boiler in the engine house consumed fourteen tonnes of wood per day.*

***43** Burra Mine in 1904, nearly three decades after closure. From left to right are the crusher house with waterwheel, the mullock dump with incline railway now removed, Morphett engine house and winding house, Peacock engine house, and mine offices. The tall wooden framework left of Morphett engine house is probably a drilling rig.*

44 Morphett chimney, engine house and winding house in 1906. Johnny Green, the miners' mascot, holding his pick and gad, stands above the engine house. The cone-shaped installation at left is a man-operated capstan used for hoisting materials from a shaft. The waterwheel at left is connected to Cornish roll crushers.

45 This derelict crushing plant, pictured in 1906, was commissioned in 1856. Water was carried along a launder, after being pumped from mine shafts, to the twelve metre overshot waterwheel which operated a ten-head stamp battery.

46 Ruins of Burra copper mine in 1906. Morphett engine house is in the centre foreground, with the open cut to the right. Behind is Peacock engine house and chimney, and the mine offices. The building in the left foreground is Morphett winding house.

47 The Big Jinker. This was built in Adelaide and used to carry large boiler cylinders from Port Adelaide to Burra. One journey in 1858 took two months, and William Woollacott, a Burra teamster, undertook the assignment for a fee of £500. The Big Jinker is now on view in the main street of Burra under care of the Burra Council.

48 Old workings in the open cut in 1980. An enormous amount of timber was needed as supports in the shafts and adits. The first contract, let in 1847, was for fifty thousand wooden logs, and subsequent contracts for up to ten thousand logs were let every few months.
During open cutting of the mine from 1870-77, old timbers were used to fuel boilers at Peacock and Morphett engine houses.

49 A southerly view of Burra copper mine and leaching plant early in 1981. Mining and stockpiling of the ore is in progress; the flat-topped ore stockpile is in the centre, with crusher and treatment plant at right. Australia's oldest intact mine building, the square powder magazine built in 1847, is encircled by a road to the immediate left of the stockpile. Part of Burra North township is in the foreground. The open cut was closed shortly after this photograph was taken.

50 The open cut in February 1981, two days before closure.

51 The cut rapidly filled with groundwater after mining ceased; this view was taken four months after dewatering pumps were removed.

52 A year later, the water has almost reached its hydrostatic level.

53 Burra treatment plant, operated by Adelaide & Wallaroo Fertilizers Ltd, in 1982. The present treatment process is more complex than the smelting of the nineteenth century. Crushed ore is roasted in a natural gas flame to produce very small metallic copper beads. After grinding, the sinter is leached with an ammonia liquor which dissolves extractable copper, and the copper-rich liquor is oxidised to precipitate chemical grade copper oxide.

The plant output of fourteen tonnes per day approximates 30 per cent of the western world's primary copper oxide production.

Kanmantoo

In 1845, the South Australian Government offered a large parcel of land near Kanmantoo to the South Australian Co. and Paringa Mining Co., both having expressed interest in several copper occurrences. The land was divided into a number of strips, and lots were drawn to determine who should have first choice, the remaining strips being divided alternately.

The more promising lodes, at Kanmantoo, were opened up by underground methods in 1846. Only ore which could be upgraded by hand picking was extracted, and sent to Swansea, Wales, for smelting and refining.

South Australia's first smelter was built on the property in 1848, together with a Cornish engine house and pump to dewater the workings.

The mine was worked by several companies, both from underground and opencut, until operations ceased in 1874 after approximately thirty-two hundred tonnes of copper had been produced.

Despite prospecting from 1906 to 1938, no further economic deposits were found. Interest was renewed in 1961 and, after numerous drill holes, a new orebody was outlined. Kanmantoo Mines Ltd was formed and overburden removal began in 1970; a concentrating plant was commissioned in 1971.

The mine had produced four million tonnes of ore containing thirty six thousand tonnes of copper by 1976 when mining ceased because of depressed copper prices.

54 Kanmantoo open cut copper mine in 1973 when the pit was in early stages of development.

55 Aerial view of the mine in 1982. The concentrating plant and offices are in the foreground, the two hundred metre deep open cut is in the middle distance, and the mullock dump is behind.

Bremer

Copper mineralisation was found by the Paringa Mining Co. near Bremer River in 1948, but lack of capital prevented installation of dewatering equipment, and mining was restricted to shallow depths.

In 1856, Bremer Mine was sold to the Worthing Mining Co., and a dewatering engine installed. Twenty thousand tonnes of copper ore were raised to 1870, when falling copper prices forced the company into liquidation.

Bremer Mining Co. worked the leases from 1872 to 1875 for a further ten thousand tonnes of ore.

56 Worthing Mining Co. operations at Bremer Mine in the late 1860s. A Cornish pumping engine was housed in the tall building at left; the building in the background housed another engine for raising ore and providing power to dressing machinery. The ore sorting shed in the left foreground partly obscures a horse whim.

Moonta-Wallaroo

The discovery of copper minerals on northern Yorke Peninsula is credited to James Boor, a shepherd; in 1859 he found green and blue carbonates in the diggings of a native rat on the site of the Wallaroo Mines. The first ore was shipped to Port Adelaide in 1860.

In 1861, another shepherd, Patrick Ryan, discovered copper minerals in a wombat hole near Moonta. He kept the find secret until reporting to the Port Wakefield publican, Mr Johnston. Two parties applied to the Registrar of Mines in Adelaide for a lease; one was organised by Johnston and the other was headed by W. W. Hughes, the pastoral lessee. Both were refused because of insufficient details of location.

The parties returned to Moonta to obtain accurate locations, and then raced back to Adelaide. W. A. Horn, riding for W. W. Hughes, travelled non-stop for twenty two hours and exhausted eight horses in the process. His application was accepted first, but a legal battle over lease ownership raged for several years, and it was not until 1868 that the courts ruled in favour of W. W. Hughes.

Moonta Mining Co. began operations at Moonta in 1862, and the first dividend was paid to share holders a year later. By 1876, dividends totalled one million pounds, an enormous sum for that period and the first time that a mining company in Australia reached this mark.

The Wallaroo Mines, at Kadina, were worked by Wallaroo Co., which also managed Wallaroo Smelters constructed on the coast in 1861. Most of the Moonta ore was also handled by these smelters.

57 Part of the Moonta mining field during 1910-20. Taylor Shaft is in the centre, with several other shafts in the distance.

Several smaller, syndicate-operated mines, including Paramatta, Kurilla, Devon Consols, and North Yelta, were successful for a time, but persistent low copper prices after 1875 forced their closure and the leases were acquired by the larger companies.

The mines brought prosperity, not only to the shareholders, but to the northern Yorke Peninsula communities and South Australia in general when mining activities at Kapunda and Burra were declining; copper production and employment in the mining industry was assured. In 1875, Moonta with a population of twelve thousand was the largest town in South Australia after Adelaide, and Wallaroo was one of the largest seaports in the State as water transport was the only means of bringing in coal for the smelters and shipping out copper. The Kadina-Port Wakefield railway connecting with Adelaide was completed in 1876.

Continuing low copper prices forced the two larger companies to amalgamate in 1889 to better use their resources. Wallaroo and Moonta Mining and Smelting Co. Ltd was now the largest mining company in South Australia, under the management of Captain H. R. Hancock and, later, his son, H. Lipson Hancock.

By 1917, the value of copper produced from the district exceeded the total for all mineral production from the rest of the State since foundation in 1836. However, activity and prosperity collapsed at the end of the First World War owing to a fall in world copper prices. Further problems were caused by limited ore reserves at Moonta Mines, and severe shortages of coal because of strikes in the eastern States. The two thousand workers at Moonta and Wallaroo refused to accept a drastic cut in wages, and in 1923 the company went into voluntary liquidation. Large stockpiles of accumulated ore and precipitates kept the smelters operating until 1926.

Moonta-Wallaroo Mines were large, even by modern standards. Mining methods were labour intensive, and use of mechanised aids such as jackpicks and air hoists did not become widespread until after 1900. Even so, the total length of all underground workings approximated one hundred and fifty kilometres. Excluding unprocessed ore shipped overseas, the smelters produced approximately three hundred and thirty four thousand tonnes of copper metal, seventeen tonnes of silver, one and a half tonnes of gold, one hundred thousand tonnes of sulphuric acid, and thirty two hundred tonnes of copper sulphate.

Small-scale mining, leaching of tailings dumps, and prospecting continued into the late 1930s, particularly at Moonta; some high grade ore pillars were mined at shallow depths.

During the 1930s, the Commonwealth and State governments subsidised mining at Moonta and a new flotation mill was installed which operated until 1938. Although a financial loss, the scheme provided employment during depression years.

Exploratory drilling by the South Australian Government from 1909 until 1949 discovered further mineralisation and, since 1960, exploration companies have located many zones of copper mineralisation, but none are of economic grade.

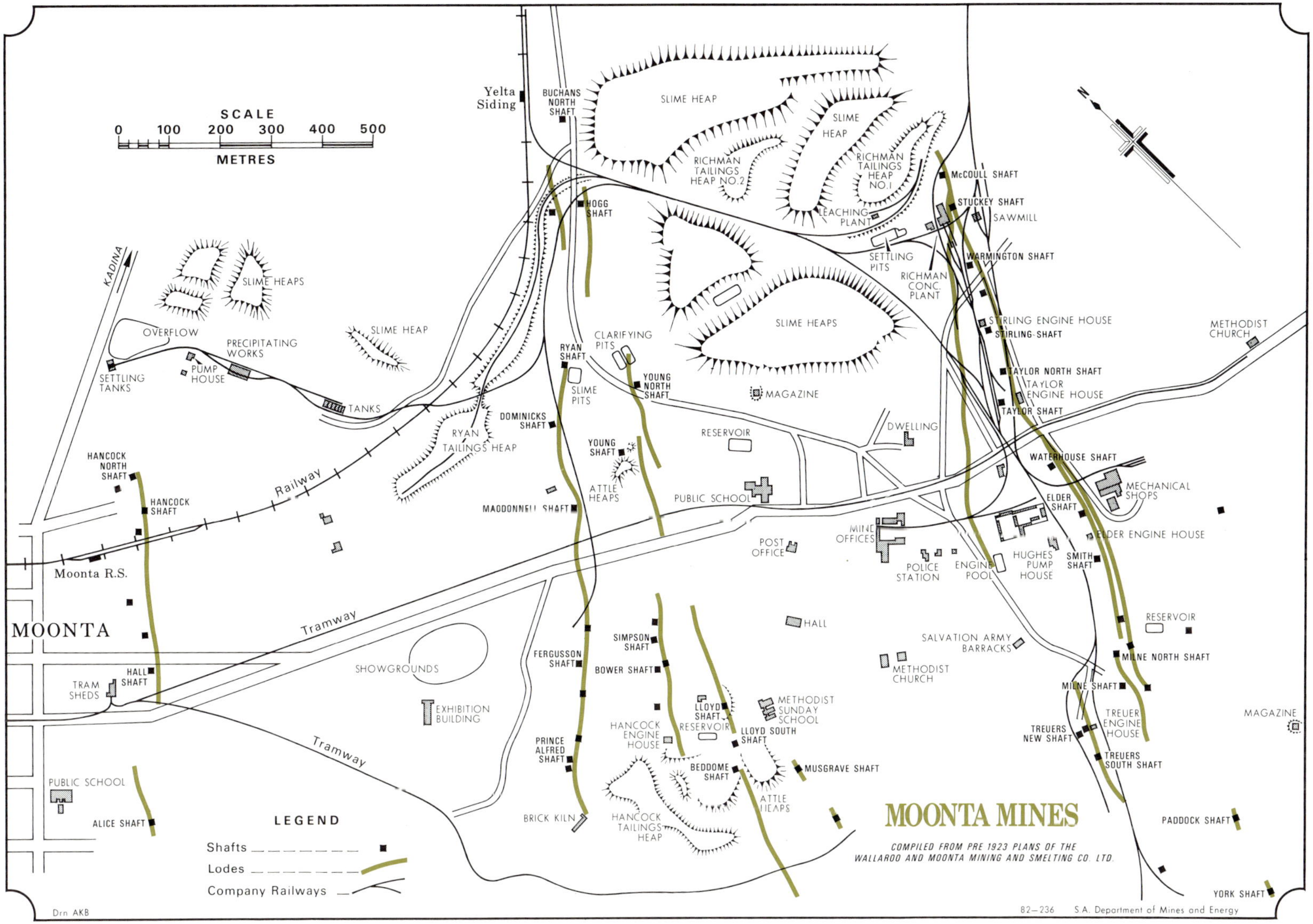
MOONTA MINES
COMPILED FROM PRE 1923 PLANS OF THE WALLAROO AND MOONTA MINING AND SMELTING CO. LTD.
SCALE
0 100 200 300 400 500
METRES
LEGEND
Shafts
Lodes
Company Railways
MOONTA
Moonta R.S.
Yelta Siding
KADINA
Railway
Tramway
Tramway
BUCHANS NORTH SHAFT
SLIME HEAP
SLIME HEAP
RICHMAN TAILINGS HEAP NO.2
RICHMAN TAILINGS HEAP NO.1
LEACHING PLANT
SETTLING PITS
RICHMAN CONC. PLANT
McCOULL SHAFT
STUCKEY SHAFT
SAWMILL
WARMINGTON SHAFT
STIRLING ENGINE HOUSE
STIRLING SHAFT
TAYLOR NORTH SHAFT
TAYLOR ENGINE HOUSE
TAYLOR SHAFT
WATERHOUSE SHAFT
MECHANICAL SHOPS
ELDER SHAFT
ELDER ENGINE HOUSE
SMITH SHAFT
HUGHES PUMP HOUSE
ENGINE POOL
METHODIST CHURCH
RESERVOIR
MILNE NORTH SHAFT
MILNE SHAFT
TREUER ENGINE HOUSE
TREUERS NEW SHAFT
TREUERS SOUTH SHAFT
MAGAZINE
PADDOCK SHAFT
YORK SHAFT
HOGG SHAFT
SLIME HEAPS
SLIME HEAPS
SLIME HEAP
OVERFLOW
PRECIPITATING WORKS
SETTLING TANKS
PUMP HOUSE
TANKS
RYAN TAILINGS HEAP
RYAN SHAFT
CLARIFYING PITS
SLIME PITS
YOUNG NORTH SHAFT
MAGAZINE
DOMINICKS SHAFT
YOUNG SHAFT
ATTLE HEAPS
RESERVOIR
DWELLING
PUBLIC SCHOOL
MADDONNELL SHAFT
HANCOCK NORTH SHAFT
HANCOCK SHAFT
MINE OFFICES
POST OFFICE
POLICE STATION
HALL
SALVATION ARMY BARRACKS
METHODIST CHURCH
SIMPSON SHAFT
BOWER SHAFT
FERGUSSON SHAFT
SHOWGROUNDS
EXHIBITION BUILDING
HALL SHAFT
TRAM SHEDS
LLOYD SHAFT
RESERVOIR
METHODIST SUNDAY SCHOOL
LLOYD SOUTH SHAFT
HANCOCK ENGINE HOUSE
BEDDOME SHAFT
MUSGRAVE SHAFT
ATTLE HEAPS
HANCOCK TAILINGS HEAP
PRINCE ALFRED SHAFT
BRICK KILN
PUBLIC SCHOOL
ALICE SHAFT
Drn AKB
82—236
S.A. Department of Mines and Energy

58 Taylor Shaft and crushing plant at Moonta Mines, circa 1918. At nearly seven hundred and seventy metres, it was the deepest Moonta shaft. In 1920, the entire engine house had to be shifted eastwards because of collapsing ground under the foundations. After 1918, all ore raised at Moonta was forwarded to this crushing plant and railed to sorting and concentrating facilities at Wallaroo Mines.

59 Richman crushing and concentrating plant, engine house, and dressing floors at Moonta Mines around the turn of the century.

60 Miners and boys from the Moonta Mines, circa 1900. Day schools, established in 1863, were poorly attended because the boys preferred to work in the mines, despite low wages. As a result, Captain Hancock established night classes for these boys and attendance was compulsory.
The Education Act of 1875 made school compulsory for children between the ages of seven and thirteen. Schools were eventually built at Moonta by the Government in 1878.

61 Blast furnaces at Yelta Mine near Moonta, circa 1905. One of the few successful independent mines, it was taken over by the South Australian Government in 1910 and operated until 1913.
The workmen are standing among wheeled ladles used to carry slag from the furnaces.

62 Miners' cottages at Moonta around the turn of the century. The township of Moonta was surveyed in 1863 and its population reached ten thousand by 1870. As most people were of Cornish ancestory, Moonta became known as 'Australia's Little Cornwall'.

Feral goats became a menace during 1860-80, and a stake fence around each property was necessary to protect home-grown vegetables and washing on the line. Naturally goat meat became an ingredient of Cornish pasties, as did rabbit, kangaroo, and wallaby

63 An 1880s view of Wallaroo Mines, at Kadina, from the top of Office Shaft head-frame.

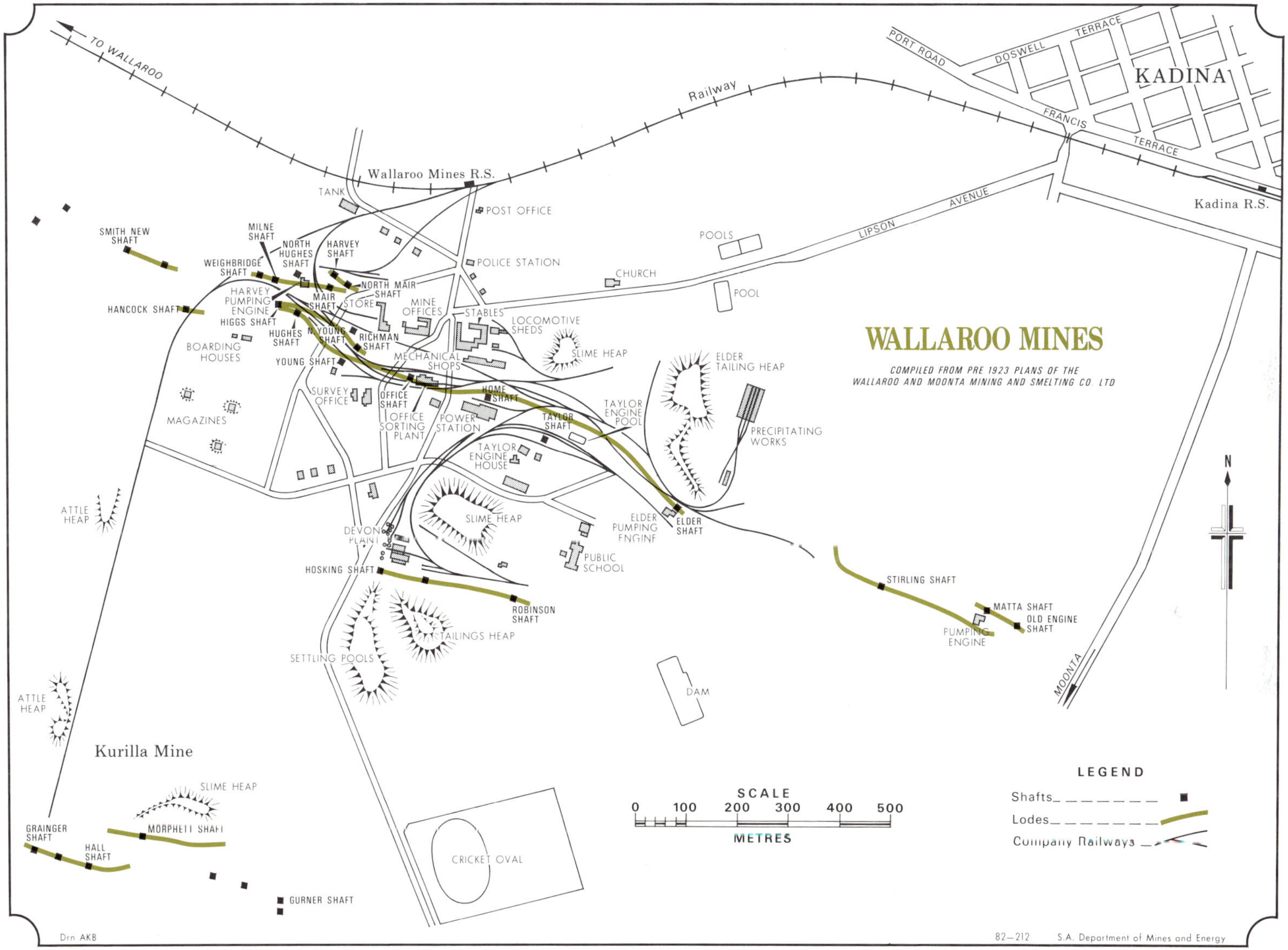
WALLAROO MINES
COMPILED FROM PRE 1923 PLANS OF THE
WALLAROO AND MOONTA MINING AND SMELTING CO. LTD
TO WALLAROO
Railway
Wallaroo Mines R.S.
KADINA
Kadina R.S.
PORT ROAD
DOSWELL TERRACE
FRANCIS TERRACE
LIPSON AVENUE
MOONTA
TANK
POST OFFICE
POLICE STATION
CHURCH
POOLS
POOL
SMITH NEW SHAFT
MILNE SHAFT
NORTH HUGHES SHAFT
HARVEY SHAFT
WEIGHBRIDGE SHAFT
NORTH MAIR SHAFT
HARVEY PUMPING ENGINE
MAIR SHAFT
STORE
MINE OFFICES
STABLES
LOCOMOTIVE SHEDS
HANCOCK SHAFT
HIGGS SHAFT
HUGHES SHAFT
N. YOUNG SHAFT
RICHMAN SHAFT
BOARDING HOUSES
YOUNG SHAFT
MECHANICAL SHOPS
SLIME HEAP
ELDER TAILING HEAP
SURVEY OFFICE
OFFICE SHAFT
HOME SHAFT
TAYLOR ENGINE POOL
MAGAZINES
OFFICE SORTING PLANT
POWER STATION
TAYLOR SHAFT
PRECIPITATING WORKS
TAYLOR ENGINE HOUSE
ATTLE HEAP
SLIME HEAP
ELDER PUMPING ENGINE
ELDER SHAFT
DEVON PLANT
PUBLIC SCHOOL
HOSKING SHAFT
STIRLING SHAFT
ROBINSON SHAFT
MATTA SHAFT
OLD ENGINE SHAFT
PUMPING ENGINE
TAILINGS HEAP
SETTLING POOLS
DAM
ATTLE HEAP
Kurilla Mine
SLIME HEAP
GRAINGER SHAFT
MORPHETT SHAFT
HALL SHAFT
GURNER SHAFT
CRICKET OVAL
SCALE
0 100 200 300 400 500
METRES
LEGEND
Shafts
Lodes
Company Railways
N
Drn AKB
82–212
S.A. Department of Mines and Energy

64 Wallaroo Mines during 1898-1910. The inset is of H. Lipson Hancock, manager of Wallaroo and Moonta Mining and Smelting Co. Ltd from 1898 to 1923, and son of Captain H. R. Hancock, the original manager.
Taylor shaft is on the left, and to its right is the white painted power station with chimney stack partly hidden by the Office Ore Sorting Plant. On the right is Office Shaft and winding house.
The upper two hundred and fifty metres of Taylor Shaft was burnt out by a major fire in 1904.

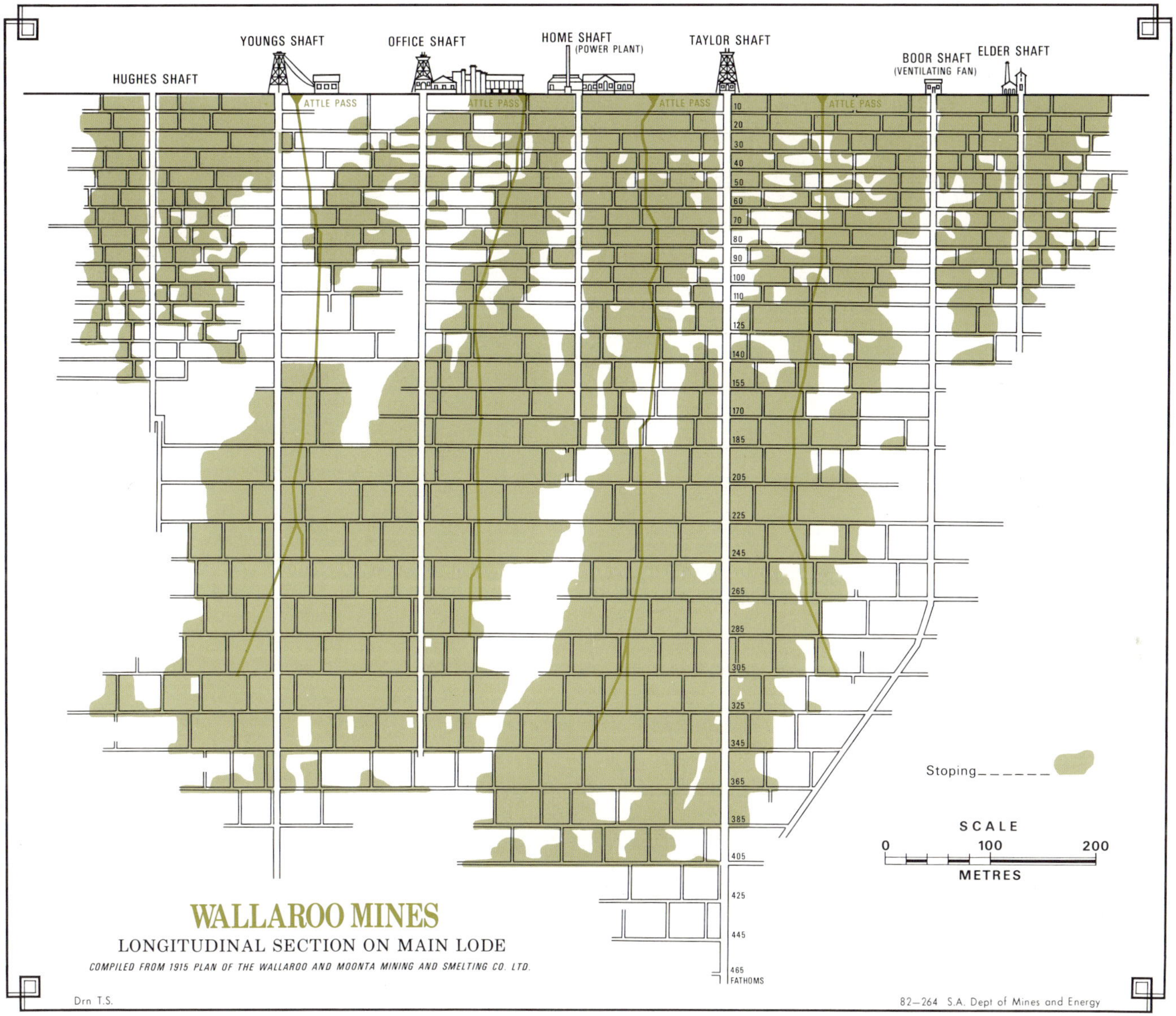

WALLAROO MINES

LONGITUDINAL SECTION ON MAIN LODE

COMPILED FROM 1915 PLAN OF THE WALLAROO AND MOONTA MINING AND SMELTING CO. LTD.

65 Drilling blastholes with jackhammers. At first, blast-hole drilling was by drill-bit and sledgehammer. A steam-driven percussion drill was designed by Captain Hancock, but exhaust steam caused intolerable heat and humidity in the mines. He later designed a new drill, driven by compressed air, which was capable of drilling twelve metres in an eight-hour shift.

66 Working a wide stope at Wallaroo Mines during 1916-18. The wooden styes were placed in position immediately after firing down ore to prevent a collapse. Stopes at Wallaroo Mines were worked by contract miners in teams of four, with an additional four labourers for transporting timbers, tools, and giving general assistance. Some of the stopes were only a metre wide. The ground at Moonta was more stable and required less timbering.

67 Miner operating a small compressed air hoist to raise ore or country rock. He is wearing a traditional Cornish miner's hat which shows white wax dripped by lighted candles mounted on the peak.

68 Large quantities of timber were required as support. In the deepest workings, even the thickest timbers were not strong enough to withstand ground swelling and, after 1902, were replaced by wooden styes filled with granulated smelter slag or waste rock.

69 Ore being tipped from trucks into haulage shoots from which kibbles in the main shaft were loaded. To the right is one of numerous ventilation doors which were opened or closed to control air flow in the mine; they were also used to confine or control mine fires by excluding air from the blaze.

70 Unloading tools and supplies from a skip at Wallaroo Mines between 1916 and 1918.

71 Ore was hauled to the surface in kibbles as at right. On the left is a man-skip, introduced to the mines in 1875. Before this, miners had to climb ladders for several hundred metres before and after their eight-hour underground shift.
Six men escaped death at Wallaroo Mines in 1894 when the wire haulage rope on a man-skip snapped. The safety catches, which had never been tested in practice, worked and saved the men from falling several hundred metres down the shaft.

72 General offices at Wallaroo Mines (date unknown). The mines employed nearly three thousand people in 1906.

73 Office Ore Sorting Plant at Wallaroo Mines in 1916. Ore was transported from the two shafts on the right by conveyor belt into the plant which had a capacity of fifty tonnes an hour.

74 Inside the ore sorting plant (date unknown). The highest grade ore was hand picked from the conveyor belt for delivery to the smelters. The remaining material was crushed and forwarded to the Devon Concentrating Plant.
A feature in the early years was the large number of boys used for hand-picking duties.

75 *Devon Concentrating Plant in 1915. The plant handled up to one thousand tonnes of ore per week. Young Shaft, with the conveyor to the Office Ore Sorting Plant, is in the middle distance.*

76 Tube mills used to grind ore in the Devon Concentrating Plant. Flint boulders were used as the grinding medium.

77 Flotation machinery was installed at the Devon Plant in 1911. Finely ground material from the tube mills was mixed with water, eucalyptus oil, and sulphuric acid in the wooden cells. Electrically driven impellers agitated the mixture with air, causing ore minerals to float to the surface on air bubbles, while waste material dropped to the bottom.
In this 1915 photograph, the silvery mineral-charged froth is flowing over the cell lip into the collection trough.

78 Aerial view of the derelict Devon Concentrating Plant in 1929. The tailings dump, with part of Kadina township, is in the background.

79 Hughes Pumping House at Moonta Mines. The engine, installed in 1865, dewatered the mines until 1923. The vertical pumping rod, made of wood, was eight hundred metres long; power was provided by a steam engine with three boilers. Cornish pumping engines were more economical than electric pumps at shallow depths, but the weight of the pumping rod made them inefficient for deep shafts.

80 Boiler explosion at Moonta Mines between 1884 and 1891. The boilers, which provided steam for dewatering pumps and other equipment, generated large pressures. A blocked valve or steampipe often resulted in a devastating explosion.

81 Electrically driven Worthington dewatering pump in Taylor Shaft at Wallaroo Mines. Taylor and Elder shafts were used to drain the entire Wallaroo mining field from 1892. Over six hundred million litres per year were raised by these pumps.

82 *Central power plant of Wallaroo Mines between 1915 and 1918. The plant generated compressed air from two gas-powered and three steam-driven engines. The air was used to operate surface and underground haulage equipment.*

83 Advances in ore treatment enabled waste material from the crushing and concentrating plants, known as 'tailings' and 'slimes', to be retreated for recovery of copper. Here, slimes at Moonta Mines are being loaded, after ploughing to enhance oxidation, into rail wagons for cartage to the leaching plant. This photograph was taken in the 1910-20 period.

84 Steam shovel loading tailings into railway wagons. The shovel was purchased for £1 425 in 1903.

85 After leaching the slimes and tailings with sea water, sulphuric acid, and iron sulphate, the copper-rich liquor was passed over scrap iron to precipitate the copper.

86 Copper precipitate is being washed to remove soluble impurities, and dried preparatory to smelting.

87 Wallaroo Smelters in 1870. These were built in 1861 and treated all Moonta and Wallaroo ore. They closed in 1926, three years after Wallaroo and Moonta Mining and Smelting Co. Ltd went into liquidation.

88 The smelters viewed from Wallaroo Jetty in 1910-20. The plume of sulphur and carbon-rich smoke was seen for kilometres, and became a well-known landmark.

89 Ore was roasted in the sintering plant to remove sulphur before smelting. Each batch was fired for about nine hours in cast-iron sintering pots on the upper level. After cooling, the pots were tipped towards the lower level to remove the sinter. Over-sized lumps were broken by hand prior to smelting.

90 Copper matte from the blast furnace was treated in the converter plant to remove remaining impurities. In the foreground is a storage furnace, containing molten matte from the blast furnace, ready to be lifted into one of two converter stands in the centre; here the matte will be reheated with fluxes to remove impurities and form blister copper.

91 *Blister copper being poured into moulds in the converter plant preparatory to railing to the refinery.*

92 Interior of the refinery at about 1915. Copper was precipitated electrolytically onto the plates while impurities, including gold and silver, dropped to the bottom as a sludge. The sludge was removed, boiled in acid, and retreated electrolytically to recover gold and silver.

93 Casting copper ingots at Wallaroo Smelters between 1902 and 1907. Hand ladling of the copper was hot, laborious work. After the ingots had been cast and solidified, they were cooled in water-filled tanks set in the floor, then carried to the weighroom for storage.

94 Weighing copper ingots. Most were sold to German brokers, but shipments went to England and India during the First World War.

Blinman

One year after copper carbonates had been found in outcrop at Blinman in 1862, four small smelters, installed by the Yudnamutana Copper Mining Co., were producing copper matte. Ore and matte were shipped to England until 1866, after which the products were sold to smelting works at Port Adelaide.

In 1872, the operating company was reorganised as Blinman Consolidated Copper Mining Co. of South Australia. Mining continued until 1874, when ore cartage costs and previous commitments to other mines forced the company to be wound up.

The opening of the railway line north from Port Augusta in 1881 provided incentive for further operations at Blinman. Several companies and groups of tributers worked the mine and treated old stockpiles of low grade ore.

A new smelter, built by Tasmanian Copper Co., was commissioned in 1904, and over two hundred and fifty men and boys were employed on site. Exhaustion of ore closed the mine in 1908.

Total production from 1862 is estimated at nine thousand eight hundred tonnes of copper metal.

95 Blinman Mine, circa 1907. The new smelter, built by Tasmanian Copper Co., is in the centre on slag dumps accumulated from the first phase of smelting which began in 1863. The square headframe of the main shaft is behind. The boiler house, with four boilers, is in the right background.

96 Similar view in 1981. Only the slag heaps and the four boilers remain.

97 Blinman Smelters in 1907. Several slag ladles are in the foreground, and the main shaft headframe is at left.

98 Similar view in 1981. Old shafts at upper left have been fenced off by the South Australian Department of Mines and Energy.

99 Blinman Mine during 1903-07. The main shaft headframe is at right, and buildings to the left house crushing and concentrating machinery. Tailings dumps are in the left foreground.

100 Similar view in 1981. The fenced-off area to the right contains the large open cut and underground workings which extend to a depth of nearly two hundred metres.

101 Tributers working in Blinman open cut during 1908-18. The main shaft and a full ore bin are in the right foreground. The three tributers are drilling a blast hole in ore by hammer and tap.

102 Blinman Mine tributers during the First World War. These miners gouged out small pockets of ore from accessible workings; they did little development work as they were only paid for the ore raised. Candles were still the only means of underground lighting and one miner has a candle tied to his waistcoat.

Yudnamutana

Yudnamutana Copper Co. was formed after discovery in 1862 of copper mineralisation at Yudnamutana. Three large blocks of ore were displayed at the International Exhibition in London in the same year but little further work was done at that time.

Transport costs were an inhibiting factor to mine development. In 1909, a Renard road train consisting of a prime mover and four trucks with a total capacity of twenty four tonnes was tested on the road between the mine and Farina Siding, a distance of about one hundred kilometres. The experiment failed owing to rough tracks and mechanical breakdowns.

A smelter was constructed on site in 1909 but no ore was treated. Copper precipitates were produced on an experimental basis from a small leaching plant in 1914.

***103** A miner's dugout near Yudnamutana Mine at about 1915. Local pastoralist/prospector W.B. Greenwood is examining a mineral specimen with a hand lens. The dugouts were not well ventilated, but provided some protection from climatic extremes experienced in the region.*

104 Yudnamutana Smelters during construction in 1909. The plant was completed in 1910 with installation of crushing and smelting machinery, but closed after a few trial runs. It was eventually demolished in 1938. Yudnamutana Mine is out of view to the right.

105 The abandoned Yudnamutana Smelters (date unknown).

Daly

Mining began at Daly, six kilometres northeast of Yudnamutana, in 1862. Worked with several other small copper deposits in the area, only the richest ore was extracted because of high transport costs.

The most active period was 1908-11 when Lyndhurst (S.A.) Copper Co. produced about fifteen hundred tonnes of ore.

106 Daly Mine during 1908-11. Only the highest grade ore was recovered, with the remainder being left in situ or dumped onto the scree slope.

Paull Consolidated

Paull Consolidated, also known as the Friend Mine, was worked from the 1890s to 1907. The main shaft was sunk to nearly one hundred metres, and an estimated thirty five thousand tonnes of ore containing one thousand tonnes of copper were raised.

107 Paull Consolidated copper mine, thirty kilometres northeast of Copley, at about 1910. The children in the goat cart are Joan and Bentley Greenwood, probably photographed during a visit by their father, W.B. Greenwood, and family for the purpose of picking copper ore from the dumps.

Sliding Rock

The mine was first worked in 1870 by a small party of miners who sold their find during the following year to the Sliding Rock Co. A shaft was sunk to nearly seventy metres, and ore containing about one thousand tonnes of copper was raised before the mine closed in 1877.

A new shaft was sunk to one hundred and ten metres by Sliding Rock Copper Pty Co. in 1899, and some ore raised. In 1906, the mine was sold to Tasmanian Copper Co. who also worked the Blinman Mine. A large dewatering pump was installed to allow mining at greater depths, but breakdown of the equipment in 1907 caused premature closure of operations.

In 1944, a pipeline was completed and water pumped from bores at Sliding Rock to Leigh Creek Coal Field until completion of Aroona Dam in 1954.

108 ***Main shaft, engine house and crusher at Sliding Rock copper mine during 1871-77. Employees in the foreground are dressing ore for smelting or transport to Port Adelaide for further processing.***

109 Sliding Rock copper mine from the now abandoned Cadnia township at about 1875.

Grunthal

Copper minerals were discovered in an outcrop near the village of Grunthal (renamed Verdun in 1917) in 1869. Several shafts were sunk and a small smelter erected. Over four hundred tonnes of copper matte were produced until a depressed metal market in 1876 forced the operations to close.

The mine was reworked for gold in 1882, but further exploration carried out in 1915 and 1935 was unsuccessful.

110 Grunthal copper mine in 1935. The Adelaide-Melbourne railway is behind the buildings.

Mount Gunson and Pernatty Lagoon

Copper was discovered near Mount Gunson nearly twenty-five years before the first recorded production in 1899, when fifty tonnes of ore were raised by the Mount Gunson Co. There was little further activity until a smelter was built in 1904; smelting was unsuccessful, however, due to high silica content of the ore, and operations ceased in 1919.

In 1906, copper carbonates were mined from Pernatty Lagoon by the Pernatty Lagoon Co. After trial leaching, a plant which could treat twenty tonnes of ore per day was built in 1915. The company went into liquidation in 1916.

From 1937 to 1943, thirty-two thousand tonnes of ore from Mount Gunson were mined and treated at the Port Pirie lead smelters.

During exploration from 1966 to 1968, several million tonnes of ore were discovered at the old Mount Gunson Mine. This was mined by Mount Gunson Mines Co. Ltd from 1969 to 1971, and a sulphide concentrate produced by a flotation plant on site. A world recession in copper prices forced the plant to be put on care and maintenance.

Further exploration located the Cattlegrid copper deposit, three kilometres west of Pernatty Lagoon. Mining began in 1974 and, to the end of 1981, over four million tonnes of ore containing seventy thousand tonnes of copper had been mined and treated at the flotation plant.

111 Copper leaching plant of Pernatty Lagoon Co. on the shores of Pernatty Lagoon in 1915. Ore was hauled up a tramway from shallow pits in the lagoon floor to a rock breaker. After grinding, the ore was roasted with sulphur and salt to produce soluble copper salts; the product was placed in vats, with salt water pumped from the lagoon, where the copper went into solution and was recovered by precipitation on scrap iron. The rock breaker is in the background, with a conveyor for transporting ore into the grinding and roasting plant. At left are the leaching and precipitating vats; a heap of scrap iron is at right. The camels were used for general haulage work, and carried copper precipitate to the nearby Birthday railway siding.

112 Construction of the Mount Gunson Mine smelter during 1904.

113 Bottrill motor tractor and wagon beside the Mount Gunson smelter in 1912. The tractor was purchased by Mount Gunson Co. to haul copper concentrates to Port Augusta. It could carry thirty tonnes and travel at three kilometres per hour.

114 Cattlegrid copper mine at Mount Gunson in 1982. The worked-out portion of the pit, backfilled with overburden and covered with topsoil, extends to the right of view from the present open cut. Native vegetation has already begun to re-establish itself in this area.

Olympic Dam

Drilling at Olympic Dam by Western Mining Corporation, in a joint venture with BP Australia, has outlined copper-gold-uranium mineralisation over an area of seven kilometres by four kilometres at depths below the surface of between three hundred and eleven hundred metres. Even at the early stages of exploration, the deposit ranks amongst the world's largest concentrations of both copper and uranium.

Evaluation and feasibility studies of the deposit are underway, and the five hundred metre deep Whenan exploration shaft is nearing completion.

115 Diamond drill rig at Olympic Dam in 1981.

116 Olympic Dam Village in 1982. The core farm, which contains most of the rock samples obtained by diamond drilling, is at right centre.

117 Whenan exploration shaft at Olympic Dam in 1982.

118 The second major lateral excavation of Whenan shaft, the four hundred and twenty metre plat, during construction. Blast holes are being drilled using air-leg-supported pneumatic drills.

119 Broken rock being mucked out of the plat.

GOLD

Although gold played an important part in the State's early development, South Australia has not been a major Australian producer.

Gold rushes, with large numbers of men hurrying to new fields, caused rapid population movements during the second half of the nineteenth century, and helped to open up the State's interior. Similar rushes in Victoria and Western Australia caused thousands of working men to leave the State, resulting in severe labour shortages. The economy was affected to such an extent that the South Australian Government passed the Bullion Act of 1852 to encourage men to return. In force for one year, this Act offered regular escorts for diggers returning with their gold from Victoria, and made gold legal tender in South Australia.

The first discovery of gold in South Australia which resulted in a productive mine was made in January 1846 at the Victoria Mine near Castambul. After this, the Mount Lofty Ranges were systematically combed, and many promising areas found.

William Chapman's chance discovery of gold at Donkey Gully near Echunga in 1852 led to the development of South Australia's first goldfield. However, Chapman received only half the government reward of £1 000 for the discovery of a payable goldfield as all conditions of the reward were not met. The remaining £500 was given to two prospectors who found the nearby, and larger, Jupiter Creek alluvial field in 1868.

Gold was panned at Forest Range in 1854 and, in 1868, a find in the Barossa Valley attracted seven thousand people to the site within six weeks. The township of Barossa, since abandoned, was established. Other significant discoveries in the Mount Lofty Ranges were at Birdwood in 1870 and Woodside in 1879.

After 1870, prospectors began to trek north and east, following the same belt of rocks that were yielding gold near Adelaide. This led to the discovery of goldfields at Waukaringa in 1873, Teetulpa in 1886, and Wadnaminga in 1888; many smaller fields were also located. Development was aided by the opening of the Silverton-Broken Hill mines in 1883, and by the extension of the railway from Peterborough to the border town of Cockburn in 1887.

Further prospecting located gold at Tarcoola in 1893, and the area developed into one of the largest goldfields in the State.

The Moonta-Wallaroo copper mines, established in 1860 and worked until 1923, yielded over one and a half tonnes of gold as a by-product.

The last major event in the history of South Australian gold was the discovery and development of the Mongolata field, east of Burra, in 1930. This, along with a rise in the price of the metal in 1931 and the economic depression, enticed many unemployed to the goldfields. Activity fell again during the Second World War.

Gold was produced as a by-product of copper mining at Kanmantoo in the 1970s and, with the increase in world gold prices since 1978, small amounts have recently been produced from many of the old mines.

The only government gold battery still operating is at Peterborough. Other batteries at Mount Torrens, Glenloth, Tarcoola, and Mongolata have been abandoned.

The total amount of gold produced in South Australia is estimated at fifteen tonnes, which is approximately 0.25 per cent of all gold produced in Australia from 1846 to the present.

The Olympic Dam copper-gold-uranium mineralisation on Roxby Downs Station holds promise for South Australia becoming a major Australian gold producer. Western Mining Corporation and BP Australia, the Roxby joint venturers, have indicated that over one thousand tonnes of the precious metal is contained in the deposit.

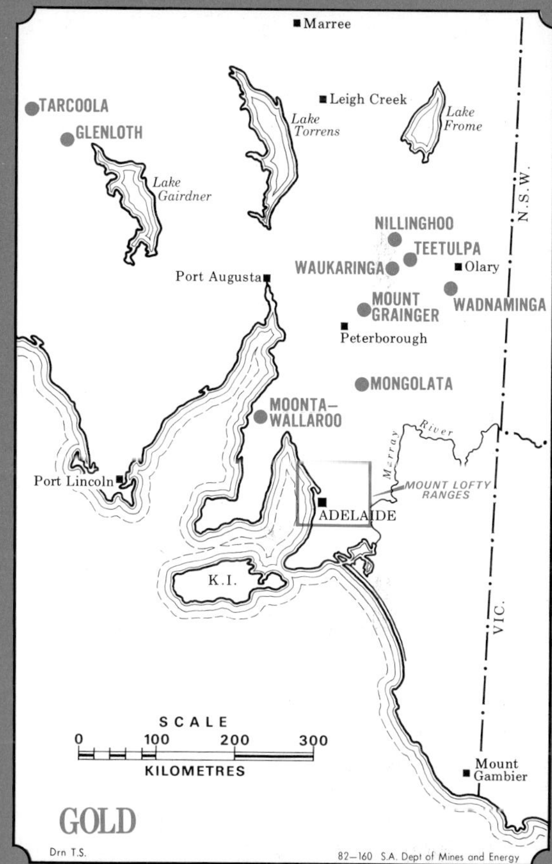
Marree
Leigh Creek
Lake Torrens
Lake Frome
TARCOOLA
GLENLOTH
Lake Gairdner
N.S.W.
NILLINGHOO
TEETULPA
WAUKARINGA
Olary
Port Augusta
MOUNT GRAINGER
WADNAMINGA
Peterborough
MONGOLATA
MOONTA–WALLAROO
Murray River
Port Lincoln
MOUNT LOFTY RANGES
ADELAIDE
K.I.
VIC.
SCALE
0
100
200
300
KILOMETRES
Mount Gambier
GOLD
Drn T.S.
82–160 S.A. Dept of Mines and Energy

Mount Lofty Ranges

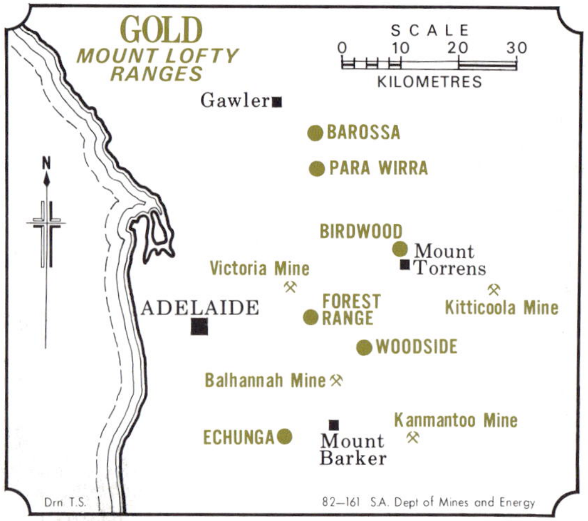

120 Forest Range Goldfield in the Mount Lofty Ranges, circa 1887. Alluvial diggings are in the foreground and miners' tents are scattered among the trees. The field reportedly produced nearly two hundred kilograms of bullion, mostly as small nuggets; the largest weighed one and a half kilograms.

121 Prospectors on the Forest Range Goldfield. Windlasses were used to lift men and dirt up the shallow shafts. Regrowth on eucalypts in the foreground indicates a recent bushfire.

122 Alluvial gold was discovered by A. Mitchell on his Woodside property in 1879; a gold bearing reef was found shortly after. The largest mine, the Bird-in-Hand, is pictured here during 1881-89. To the left is the winding house, and a tramway for cartage of ore links the shaft with the battery house in the middle distance.

123 The twenty-head stamp battery at Bird-in-Hand. A steam-driven axle with cams raised and dropped the heavy iron stamps, while gold ore was fed into the crushing boxes below. Gold was recovered from the resultant powdered rock by amalgamation with mercury on copper plates.

124 The Bird-in-Hand was reopened in 1934. The mine was dewatered by an electric pump, and this water is flowing from the launder in the foreground. In the background is the old pump house which formerly housed a Cornish pumping engine.

125 Employees beside the main shaft during dewatering operations in 1934.

126 Department of Mines diamond drilling plant at Bird-in-Hand Mine in 1934.

127 Nest Egg Mine at Woodside in the late 1880s. Sinking of the main shaft on the right was suspended after a large inflow of water.

128 Northerly view of Ridge Gold Mine near Woodside before 1887. The battery house is in the centre, and white sand tailings produced from crushed ore lie in the creek. The area has been almost cleared of trees to provide fuel for boilers.

129 Stamp battery house at Two-in-the-Bush Mine, Woodside, during 1881-89. Ore entered the battery along the ramp at the rear of the building. The trench in the foreground carried away waste water and tailings.

130 Main shaft headframe and buildings of Balhannah Mine in 1907. The buildings housed an engine and boiler for raising and lowering the ore skips, and a roll crusher.

131 Grinding ore with a roll crusher at Balhannah Mine during 1867-76. Lump ore shovelled beneath the revolving millstones was ground to a fine powder suitable for gold extraction. The mine also produced copper and bismuth.

132 Alluvial gold was discovered at Echunga by William Chapman in 1852 and South Australia's first goldfield was created. By 1907, this water sluice had been erected to rework the large Jupiter Creek alluvial goldfield, discovered in 1868. Alluvial gravel was placed at the top of the sluice and washed over corrugated riffles where gold collected. Only small quantities were recovered, however, and operations stopped after eighteen months.

133 The shaft headframe, crushing and treatment plant of Para Wirra Mine, south of the Barossa Goldfield, in 1939.

134 An easterly view across Reedy Creek of the Kitticoola Mine near Palmer, circa 1900. The bifurcated flue at right, and stone-work in the centre, are the remains of a copper smelter operated about 1850 by the Australian Mining Co. The company may not have realised that the ore treated also contained gold. An adit which provided access to the main lode is almost at creek level below the large boiler.

135 A westerly view of Kitticoola Mine in 1908. A shaft is in the left foreground and an open cut is in the centre. Steam is emerging from wood-fired boilers which powered an air compressor to operate mine equipment.

136 A similar view in 1935 showing the recently erected battery house.

137 The ten-head stamp battery operating at Kitticoola in 1935. The mine produced over one hundred and fifty kilograms of gold bullion.

Waukaringa

Reef gold was discovered to the north of Yunta by a shepherd, J. Watson, in 1873 and several mines were opened. The largest was the Alma and Victoria, which produced nearly one and a half tonnes of bullion from over two kilometres of underground workings before closure in 1894.

Many miners and their families were attracted to the area, and the township of Waukaringa was proclaimed on 1 November 1888. The town provided facilities such as school, post office, hospital, hotel, butcher and police station until it was abandoned in the 1950s.

***138** White's Lovely Gully Mine near Waukaringa in 1930. Lovely Gully was first worked between 1891 and 1895. The economic depression of the 1930s led to the field being re-evaluated, but little further mining was undertaken.*

***139** The Alma and Victoria Mine was the largest on Waukaringa Goldfield. It had twelve shafts, the deepest being over three hundred metres deep, and a twenty-head stamp battery. The mine closed in 1894 owing to a lowering of ore grades, but the tailings dumps were retreated by cyanidation in 1899.*
A steam engine ran the dewatering and crushing machinery, and this photograph taken in 1975 shows the ruins of the engine's chimney and flue.

Teetulpa

Nugget gold was discovered a few kilometres north of Yunta by Thomas Brady and Thomas Smith in October 1886; five thousand prospectors rushed to the field within two months.

Gold was prised from gutters and crevices of the country rock, and panned from alluvial leads of fossil stream beds. The largest recorded nugget weighed almost one kilogram.

Mining was hindered by the lack of water and, at first, all alluvial ore was carted three kilometres to Tonkin's Well. Later, water was pumped to tanks at the southern end of Brady's Gully, which enabled miners to wash larger amounts of pay-dirt.

By 1889 the field was virtually deserted, although small-scale mining has taken place since. Total estimated production was nearly three tonnes of bullion.

140 Brady's Claim, discovery site of the Teetulpa Goldfield, in late 1886.

141 Alluvial diggings on Windlass Hill in the period 1886-88. This area of high ground had good wash material up to eight metres thick, which allowed the prospectors to treat large volumes of alluvium.

142 and 143 Washing gold from alluvium with gold cradles at Tonkin's Well in the period 1886-88. Alluvium was shovelled onto a sieve at the top of the cradle and washed with water while the apparatus was rocked from side to side. The heavy fraction which collected in a series of riffles was panned to recover the gold.

144 Prospectors panning for gold. All water for washing on site had to be hand carted and, consequently, the water was used several times before being discarded as too dirty.

145 Dipping water for gold washing from tanks at Brady's Gully.

146 A goldfield's warden settling a dispute. He was required to decide quickly and fairly on matters ranging from claim jumping to theft to maintain relative calm among the several thousand prospectors.

settling a dispute

147 Most miners lived in canvas tents, although some sheet iron shanties were erected.

148 The gold miners lived a harsh and sometimes unruly life, but many regularly attended church services or sought counselling from the chaplain. This photograph shows the Wesleyan Chapel.

Mount Grainger

Gold was assayed in silver ore from the Buttamuck Mine in 1889, and marked the beginning of one of the State's smaller goldfields. Several gold-bearing quartz-ironstone reefs were found, and companies formed to work them.

The largest mine was Mount Grainger, which produced fifty kilograms of bullion from 1894 to 1915. Alluvial mining yielded gold from a creek north of the mine.

Some claims near Mount Grainger are still worked.

149 Prospector operating a dry blower on the Mount Grainger Goldfield, northeast of Peterborough, in 1930. The dry blower is used on alluvial fields where water is scarce. Dirt is shovelled onto the upper shaking screen through which finer material passes to a riffle on the centre tray. Here the gold and other heavy minerals are separated from lighter waste by puffs of air forced through the riffle by the bellows underneath. At regular intervals the concentrate is removed and panned to recover the gold. The entire operation is controlled by foot and hand levers.

Wadnaminga

Reef gold was found at Wadnaminga in 1888 by three prospectors who named their claim Eiffel Tower. Other mines were developed along the reef, and included Victoria Tower, Oulnina Tower, and Countess of Jersey.

In 1889, another line of gold-bearing lodes, The Birthday, was discovered northeast of Eiffel Tower. Mines opened on this line included Great Eastern and Virginia.

The Eiffel Tower was characterised by rich pods of gold, which contrasted with The Birthday where the gold was finer grained and more evenly distributed.

Approximately four hundred miners and three ten-head stamp batteries were on the field in 1892. Mining was almost continuous until 1897, but there were few miners left by the end of the First World War and the field was finally abandoned in 1936, having produced over a half tonne of bullion.

***150** Virginia Mine in 1915. A chute from the battery house carried tailings to the two settling tanks which are directly behind the ore truck. The tailings were then transferred to the cyanidation vats in the left foreground for final gold extraction.*

***151** Assay office at Virginia Mine. The building which was constructed of local stone during 1850-95 was reopened in 1912, but closed again in 1915.*

152 Ten-head stamp battery during construction at Great Eastern Mine on Wadnaminga Goldfield in 1915. The battery was erected in an attempt to reopen the mine that was first worked from 1889 to 1893. The stamps were imported, but the driving wheel was manufactured at the Wadnaminga forge. The machine was used for less than two years, and was advertised for sale in 1919.

153 The suction gas engine which powered the Great Eastern stamp battery. The engine was fuelled by locally produced wood gas.

Tarcoola

Alluvial gold was discovered by a local shed hand, Nichols, on the eastern end of Tarcoola Hill in 1893. This was followed in 1900 by Fabian and Ward's discovery of reef gold in the near vicinity. The Tarcoola Blocks Mine, developed on this lode, became the largest gold mine in the State. From 1904 until closure in 1912, the mine employed one hundred and seventy people, had a twenty-head stamp battery, and a cyanidation plant to treat battery tailings.

Smaller mines on the field were Tarcoola Perseverance, Curdnatta, Royal George, and Welcome Home.

Total estimated production from the field is over three tonnes of bullion, and includes small amounts won in recent years.

***154** Tarcoola Blocks Mine, circa 1930. To the right are a flue, chimney, and water cooling towers to conserve water from the boilers in the foreground. In the centre background is Fabian's Shaft, site of the original gold discovery, and at the extreme left is the stamp battery.*

***155** Government gold battery and cyanide works at Tarcoola. The ten-head stamps began operating in 1901 and closed in 1955. The battery helped to keep smaller mines of the district viable.*

Nillinghoo

Nillinghoo is another of the State's smaller goldfields, and its only significant mine is Kirkeek's Treasure. This has produced ninety kilograms of gold bullion since Henry Kirkeek's find in 1894. The mine is still worked.

156 Kirkeek's Treasure Mine on Nillinghoo Goldfield, north of Teetulpa, circa 1907. The tramway carried ore to the ten-head stamp battery for crushing.

Mongolata

Henry Byles, a drover, discovered a gold bearing quartz reef at Mongolata in 1930. This helped to stimulate a renewed interest in gold mining in South Australia over the next few years, and prompted the government, with assistance from a Commonwealth unemployment relief loan, to build a battery and cyanidation plant on the field. This was officially opened in 1933 by R. S. Richards, the South Australian Premier.

157 Byles gold mine at Mongolata in 1931. A covered stockpile of hand picked, bagged gold ore awaits transport to the battery. The mine was the largest producer on the field, yielding over one hundred kilograms of bullion before closure in 1949.

158 Easterly view of Mongolata Goldfield in 1933. The government battery and cyanidation works are at right centre.

159 Entrance adit to Takati gold mine in the late 1930s. The owner, William Pexton, is panning for gold with his sons. The mine was worked from 1932 to 1942, and was noted for large pieces of specimen gold, the largest recorded weighing nearly seven kilograms.

AGGREGATE

Crushed Rock

Aggregate was first quarried in South Australia near Kingscote in 1836, and the naturally jointed basalt was taken to build the road between Port Adelaide and Adelaide.

Quarries were opened at Stonyfell by James Edlin in 1837, and at Beaumont by the Greenhill Slate and Flag Quarry Co. in 1838. Others were established along the foothills of the Mount Lofty Ranges to provide hand-hewn building stone and flagstone.

Adelaide roads were paved at first with local creek gravels; later, hand-broken stone for road construction was derived from quarries at Yatala Labour Prison and Stonyfell.

Quarrying methods were both labour intensive and dangerous. Small ledges, barely wide enough for several men to stand, were cut into sheer quarry faces, in some cases exceeding one hundred metres in height. Holes were drilled into the rock by teams of three men using 'hammer and tap': one man held and rotated a steel rod which was struck alternately by two men wielding sledge hammers. The holes were loaded with gunpowder and fired to bring down stone which was broken on the quarry floor with hand tools. Rocks too large for breaking by hand were drilled and blasted at the quarry face. Steam-driven crushing plants were installed before the end of the nineteenth century, and rail trucks were used to cart stone from the face to the crushers.

Although new explosives and equipment such as pneumatic drills and face shovels became available, basic quarrying techniques changed little until after the Second World War. Improved blasting practice, and new regulations introduced under the Mines and Works Inspection Act in 1954, reduced the danger of serious rock falls by limiting face heights to twenty metres. This resulted in quarries with several benches, effectively increasing their working areas and hence production.

The cost of quarry products to the community has declined since the early days. In 1855, the *Government Gazette* advertised road metal from the newly opened Dry Creek quarries at Yatala for six shillings per cubic yard, which approximated the daily wage for a labourer at that time. A similar product today costs about eight dollars, substantially less than a day's wage.

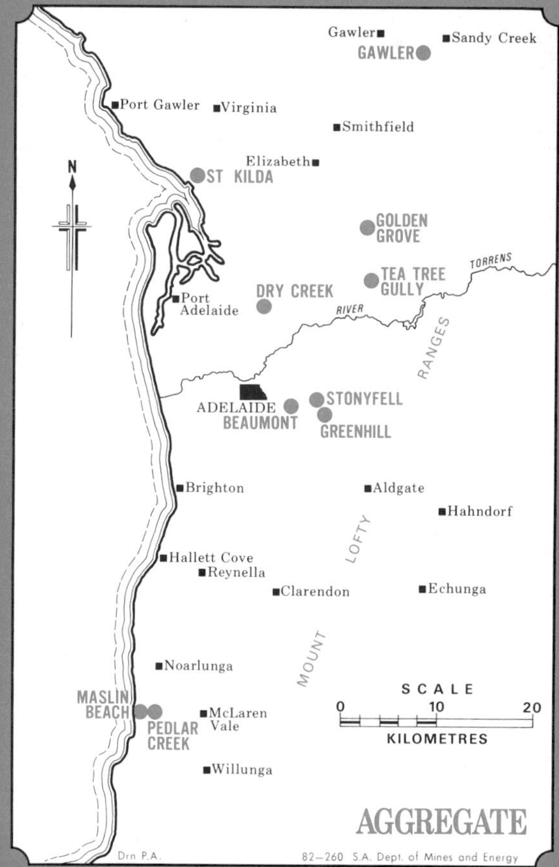

160 Horse dray being loaded with hand-broken stone in the Kapunda district during the late 1800s.

161 Dry Creek stone quarries in the grounds of Yatala Labour Prison during the late 1800s. An inclined screen for sizing broken rock is in the left foreground, and several stockpiles of graded material are to the right. This quarry has since been filled in and landscaped.

162 Quarrying at Tea Tree Gully during the early 1900s. A large boulder which has fallen from the quarry face is being prepared for secondary blasting. Three workers in the foreground are drilling a horizontal hole for explosives using 'hammer and tap'. A compressed air jackhammer is being used to drill a hole in the top of the boulder.

163 Stonyfell Quarry in 1921. The crushing plant is in the foreground and the high quarry face behind. Most of the hill on the left has now been removed.

165 Carting broken quartzite to the crusher.

164 Quartzite blasted down to the quarry floor at Stonyfell is being broken with a spalling hammer and loaded into rail trucks.

Sand and Gravel

Deposits of construction sand and gravel were first worked along the banks of the River Torrens. Simple hand mining methods were used initially, but were replaced later by horse-drawn scoops. Reserves of natural gravel were small, and a preference by builders for the cleaner, more uniform crushed aggregate saw most gravel pits closed by the end of the nineteenth century.

For many years, deposits between Gilberton and Paradise were the principal source of construction sand for metropolitan Adelaide. These deposits were depleted by the 1940s, and St. Kilda became an important source.

Sand mining began in the coastal cliffs at Maslin Beach in 1928 with The Noarlunga Sand Co. Ltd, which erected an aerial ropeway to convey the product to a railway siding three kilometres northeast. This deposit has been worked continuously to the present time under several companies, with output totalling over eight million tonnes. Three other major pits have operated nearby since the 1960s.

As a consequence of greatly expanded requirements to serve the metropolitan area after the Second World War, the extensive sand deposits at Gawler and Golden Grove were opened up.

A variety of extraction techniques have been used to recover sand, including draglines, face shovels, and barge-mounted hydraulic monitors. Sand is now generally won by scrapers loading from inclined pit floors to blend the horizontal layers. All of the major pits have installed washing plants for the removal of clay and production of a variety of sand grades.

166 Sand pit near Gawler in 1965. An excavator (left) and dragline (right background) are loading dump trucks.

167 An elevating scraper loading sand from the inclined floor of a sand pit at Gawler in 1974. The elevating scrapers not only carry large tonnages, but are quick to load and can blend sand from several levels to provide a relatively uniform product.

168 A sand washing and grading plant at Pedlar Creek. After washing and sizing, the sand is conveyed by radial stackers to the conical heaps for drying.

169 Sand, clay, and brickshale pits at Golden Grove in 1980. Stonyfell Quarry is the prominent white landmark in the centre background.

Rehabilitation

In September 1921, *The Observer* featured a photograph of Stonyfell Quarry with the caption 'picturesque view of a quarry'. By the early 1970s, however, public opinion had changed considerably, and an increasing awareness of environmental impairment brought significant changes to the quarrying and sand mining industries.

The Extractive Areas Rehabilitation Fund, administered by the Department of Mines and Energy, was set up in 1972. Financed by a levy on sand, rock and clay produced, the fund is used for rehabilitation of areas disturbed by quarrying, and finances research into methods of reducing environmental changes caused by quarry operations.

The largest rehabilitation project undertaken to date was based on the eastern faces of Stonyfell Quarry. When completed, this will significantly reduce the visual impact of a prominent Adelaide landmark.

170 Stonyfell Quarry and crushing plant in 1956. The high working face has now been replaced by a series of benches which are safer, and more suited to modern quarrying practice.

171 Stonyfell Quarry in 1980. The benches at left have reached their maximum extent and preparations are being made for rehabilitation.

172 A similar view in 1982. Benches on the left have been covered with mullock and sown with grasses prior to planting of native trees. Approximately thirty million tonnes of quartzite have been quarried at Stonyfell since 1837.
On the right is the adjoining Greenhill Quarry.

173 Stonyfell Quarry and a portion of Adelaide in 1982. Backfilled and rehabilitated benches are in the foreground, and the present quarry is behind. Native trees and shrubs have been planted in the thick grass matting which was sown to rapidly stabilise the topsoil and provide a mulch for the young seedlings.

174 ACI Operations Pty Ltd glass sand pit at Normanville. The working face is in the background, and progressively rehabilitated land in the foreground and to the right. Mining has removed the centre of the dune, leaving natural slopes on the seaward and landward sides of the pit.

DIMENSION STONE

Natural stone has been an important construction material in South Australia because of a scarcity of suitable natural timber. Not only were public buildings, churches, and large private offices built of stone, but it was commonly used in dwellings.

The earliest buildings in Adelaide were constructed of the nearest available materials including River Torrens gravel and surface limestone, or calcrete, that occurs throughout the North Adelaide and Prospect districts. Calcrete became so popular that people began to quarry in parklands and road reserves to such an extent that the Government was forced to ban the practice. Calcrete was also quarried near many country towns such as Cowell and Port Wakefield.

Bluestone was first quarried from the Adelaide foothills in 1838, and was the dominant building stone for houses until the turn of the century. Quarries were operated from Magill to Mitcham, with the principal output coming from Glen Osmond; the Tapley Hill quarries became important producers in the 1870s. Bluestone was also mined from underground on a small scale at Magill. Many country towns including Clare, Auburn and Gawler used a similar material.

Quarrying into the banks of the River Torrens on the present sites of Torrens Parade Ground, Adelaide Festival Centre, and Adelaide Railway Station, produced fossiliferous limestone. Many important early buildings were constructed of this stone, one example being the Old Legislative Council Building now refurbished as the Constitutional Museum.

During the 1860s and 70s, attention turned to high quality sandstones in the Adelaide Hills, particularly those at Tea Tree Gully. Prominent public buildings such as the Supreme Court, General Post Office, and Adelaide Town Hall are constructed of Glen Ewen and Tea Tree Gully sandstones, which are among the best used in Australia.

In the latter half of the nineteenth century, marble was quarried and used locally at Kapunda, Angaston and Macclesfield. The first major use of marble and granite in South Australia was during 1883-89 when the western third of Parliament House was constructed using Kapunda marble on a plinth of granite from West Island, Encounter Bay. Fifty years later the remainder of the building was completed in the same materials.

Economic depression in the 1890s, and rising costs of working hard sandstone, forced diversification towards softer and more easily quarried and dressed stone. Limestone from the river cliffs at Murray Bridge, and later at Ramco and Waikerie, gradually replaced Tea Tree Gully sandstone as the principal stonework of prominent buildings.

Granite from Monarto and Murray Bridge provided stone for buildings, street kerbing and statue plinths in the early part of this century.

Glen Osmond and Tapley Hill bluestone were progressively replaced in the early twentieth century by sandstone from Mount Lofty, Stirling, and Sheoak Hill. Since 1960, the trend has continued towards softer material, and Basket

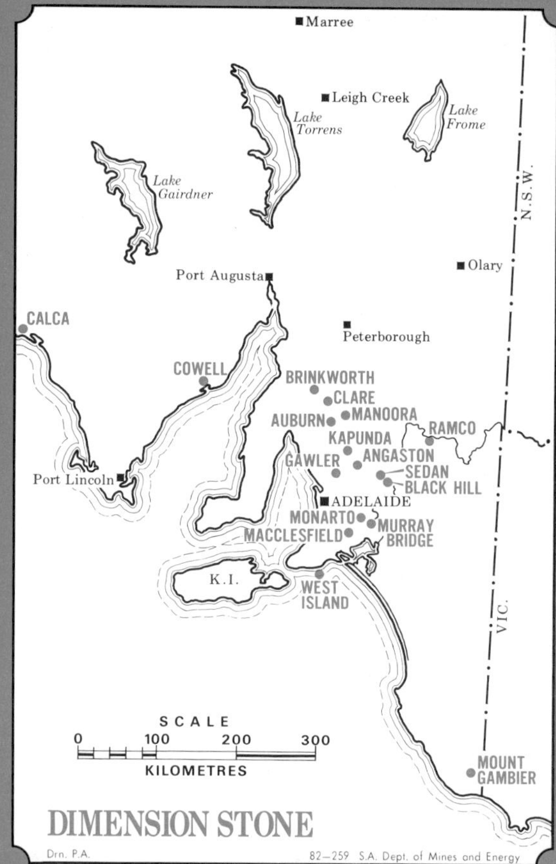

Range sandstone, popular in the 1950s, has been replaced by stone from Kapunda, Brinkworth and Manoora. However, these quarries have now closed and little sandstone is currently worked for house construction. The only significant source is a quarry at Basket Range.

As a result of a shortage of building materials after the Second World War, considerable quantities of limestone were brought to Adelaide from quarries at Mount Gambier; this limestone was first used in 1844. Six quarries currently produce the light, porous, and readily sawn stone.

With the advent of steel and concrete frames, stone ceased to be a structural element in the construction of multi-storey buildings and is now used as a veneer or cladding to curtain walls supported at each floor by a structural skeleton. Thin slabs of polished granite and marble have thus replaced solid blocks.

Today, the use of stone is relatively small, as the natural product faces considerable competition from exposed aggregate and other concrete finishes and stone substitutes. Consequently, the South Australian stone industry is principally concerned with granite and marble for the monumental trade and for occasional building work. Marble is being worked at Angaston, and granite at Calca near Streaky Bay, and at Sedan. Two quarries are operated at Black Hill, where large quantities of high quality black norite have been worked since 1958.

175 Angaston marble quarry in 1921. Lines on the marble blocks are drill holes in which 'plug and feathers' were driven to split the stone. The quarry is still worked today, but for coloured aggregates rather than building stone.

176 A limestone quarry at Mount Gambier in the early 1920s. The stone was hand sawn into pillars about one metre square, then toppled to the quarry floor where it was sawn either directly into building blocks or into large cubes for transport to the building site where it was worked as required.

177 Granite quarry at Murray Bridge in the early 1920s. Quarrymen are producing large regular blocks with a minimum of drilling and splitting. Though relatively small, this quarry was one of four or five that produced Murray Bridge granite.

178 Hand sawing of a toppled pillar of Mount Gambier limestone in 1950. Note the solution cavities and root holes in the fossiliferous limestone.

179 Modern power-driven circular saw undercutting limestone blocks, that have already been cut in two vertical planes, to directly produce building blocks known as ashlars.

180 A view over some of the extensive quarrying operations at Mount Gambier in 1973.

181 The largest single granite quarry in Australia; this is one of two quarries at Black Hill near Mannum which produce black granite, more properly known as norite. Blocks are worked by quarry bar mounted rock drills, and hoisted by the fifty tonne capacity crane.

182 Modern quarrying methods at Black Hill in 1981. Compressed air driven rock drills, supported on quarry bars, drill both vertical and horizontal holes around the block being quarried. In this case the stone splits so poorly that the vertical walls are completely drilled free; light explosive charges are used to free the base of the block.

CLAY AND SHALE

The absence of timber suitable for building purposes in Adelaide induced brick manufacturing to start soon after establishment of the Colony in 1836.

Eight brickmakers were operating by 1838, using red alluvial plastic clay from the banks of the River Torrens. Two of the earliest manufacturers were East and Breeze, who operated near the Morphett Street bridge, and J. F. Aldenhoven who worked at Underdale. Several plastic clay sources along the foothills of the Mount Lofty Ranges were also exploited, particularly at the Gilburn brickworks.

Pug mills to mix raw materials were operated by horse and bullock, but brick forming was done by hand until 1840 when steam-powered plants capable of extruding twenty thousand bricks per day were introduced. Even so, horse-powered plants were still in general use until the early twentieth century.

Kilns were of the intermittent firing type until the first Hoffman kiln was built at Blackwood in the 1880s to supply bricks for lining tunnels along the hills' railway line. The first tunnel kiln was erected by PGH Ceramics at Renown Park in 1961.

Underground mining of white plastic clay began at Tea Tree Gully in 1880 following parliamentary endorsement in 1878 for the construction of a sewerage system in Adelaide and suburbs, and the consequent need for large quantities of fired clay pipes. Operations were underground until 1962, and open cut until the early 1970s.

Although many types of commercial pottery were manufactured in nineteenth century Adelaide, few pottery houses persisted. J. C. Koster established his works at Beulah Park in 1880, and became a major supplier of electric insulators and jugs. The pottery closed in the mid-1970s and was purchased by the Payneham City Council in December 1981. Bennett's pottery at Magill, founded in 1888, is still in operation producing a variety of terra cotta and stoneware vessels.

Clay roofing tiles were introduced to Adelaide in 1910 by Wunderlich Ltd of Sydney, who opened a branch office to sell imported tiles from France. By 1920, the company was manufacturing tiles at Edwardstown, and the plant is still in operation using plastic clay from Cherry Gardens.

During 1919-26, a period of marked expansion in the Adelaide building industry, many clay and shale pits were opened; brick production was aided by the introduction of electricity in 1925. Manufacturers who started at this time were the S.A. Co-operative Brick Co. on red plastic clay at Brompton, Federal Brickworks on red plastic clay at West Thebarton, and City Bricks Ltd on shale at Glen Osmond.

By the late 1950s, the traditional clay sources along the River Torrens were nearing depletion. Two major brickmaking companies, Hallett Brick Industries Ltd and PGH Ceramics, established new plants on clay and shale deposits at Golden Grove. They are now the primary suppliers of bricks to metropolitan Adelaide, supplemented with production from the old plants near the River Torrens and at Glen Osmond, and a new plant at Lonsdale.

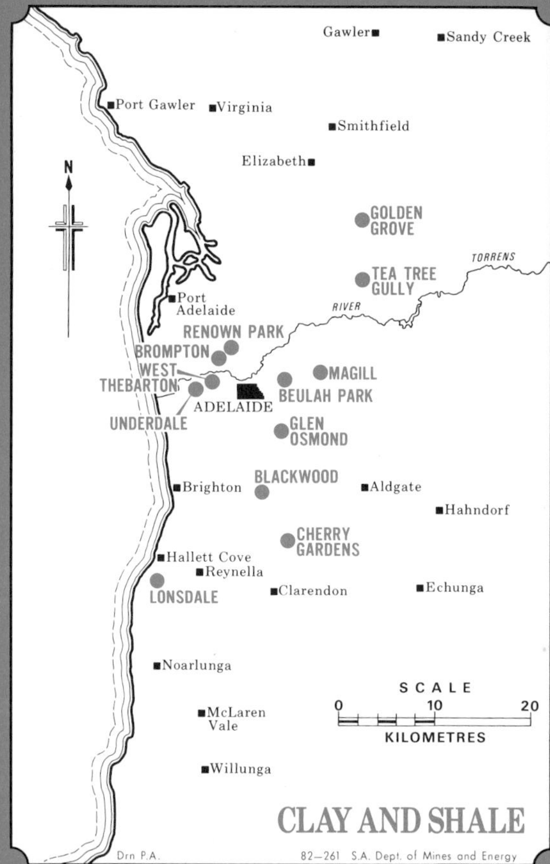

183 Plant and Hoffman kiln of the Metropolitan Brick Co. at Magill, circa 1922. Clay had been mined from this site since 1840. Gilburn Brick Co. operated the plant from the mid 1930s until purchased by Hallett Brick Industries Ltd in 1974. In 1975, seven hundred thousand pressed bricks were manufactured for the paving of Rundle Mall, using shale and clay from Reynella, Golden Grove and Anstey Hill, as well as the Gilburn deposit. The pressing plant was shifted to Golden Grove at the end of 1981 to utilise the larger kiln capacity there, and the brickworks have since closed. The average weekly production before closure was two hundred and fifty thousand paving and house bricks.

184 Loading of white plastic clay at the headframe of Lambert and Parry clay mine near Tea Tree Gully in 1923. Underground mining began in 1880 to supply raw material for sewerage pipes and pottery, and later for bricks. The deposit was purchased in 1953 by Adelaide Potteries, and operations converted to open cut. Christies Sands Pty Ltd bought the leases in 1964, worked the clay, and established a washing plant to treat the overlying sand. The deposit is still held by Christies Sands for white plastic clay and coloured shale. The company has been part of Hallett Brick Industries Ltd since 1965.

185 Poppet head, loading platform and mine timbers at the Adelaide Pottery and Drainpipe clay mine near Tea Tree Gully in December 1923. The company started operations in 1878, although the white plastic clay had been worked since 1849. Underground mining continued until 1962 when PGH bought the operations and open cut the deposit. Christies Sands Pty Ltd purchased the leases in 1964, but the land was sold to the Tea Tree Gully City Council later that year for use as a rubbish dump.

186 S.A. Co-operative Brick Co. clay pit in Renown Park at about 1956. The deposit was first worked in 1921. The dark coloured pit face at left comprises in situ red clay, whereas the white material to the right is stockpiled weathered shale from Houghton. The two materials were transferred by rail truck to the plant (out of view) for batching and firing. PGH operated the pit from 1959 until closure in 1970, although the brick kilns continued production, utilising imported clay and shale, until 1977.

187 Federal Brickwork's red alluvial clay pit in West Thebarton at about 1952. Hallett Brick Industries Ltd purchased the operations in 1910, installed a Hoffman kiln in 1911, and worked the pit until closure in 1977. The pit was subsequently purchased by the Thebarton Corporation and partially filled in; it is now being developed as a recreation area.

188 City Bricks Ltd shale quarry at Glen Osmond, circa 1953. Mining began at about 1924 and continues to the present day, although PGH have been the operators since 1971. Rock was hauled from the quarry in railcars; tyred trucks are now used. After crushing, the shale is blended at the brick-works with clays from other metropolitan deposits and fired. Present production is twenty million bricks per year.

SLATE

Slate deposits are widespread in South Australia, and have been exploited at a number of localities in the Mount Lofty and Flinders Ranges.

Those at Willunga and Mintaro are Australia's most historically significant slate quarries. They produced complementary materials: Willunga slate is thinly cleaved and thus suitable for roofing, while Mintaro slate is thicker bedded and ideal for paving and structural applications.

In addition to these, early working of slate in South Australia included small quarries at Golden Grove, Chinkford (now Manoora), and Andrews near Spalding. More recently, deposits near Manna Hill, Jones Hill, Kanmantoo and Wistow have been investigated.

Since 1968, significant quantities of paving and walling stone have been produced from the Wistow quarries. With a slightly rough surface and irregular shape, the Wistow stone has a rustic appearance and is used in a variety of domestic applications.

Increasing demand for slate flooring tiles, particularly with variegated colours, has stimulated production from the Jones Hill Quarry, at which operations began in 1976. Apart from the coloured surfaces, Jones Hill slate is similar in most respects to the smooth, blue-grey slate that is still worked at Mintaro.

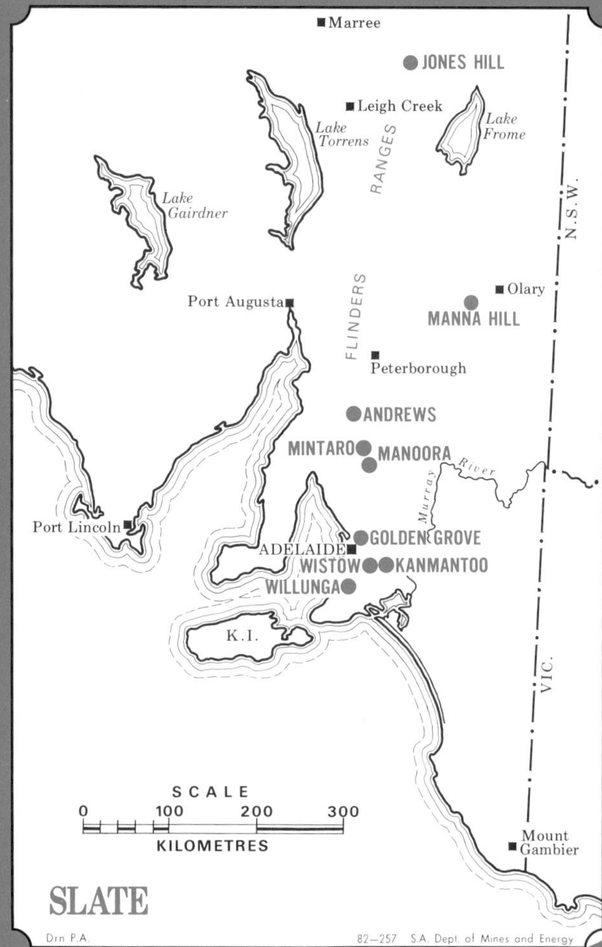

Willunga

Slate suitable for splitting into roofing shingles was discovered by Edward Loud near Willunga in 1840. By 1841, five separate quarries were in operation; the four important producers, all established by 1860, were Bangor, Delabole, Martin's and Bastian's.

Slate shingles were carted from the quarries by bullock dray to Aldinga, and later Port Willunga, for shipment. The industry reached a peak in 1891 when two vessels, each carrying twenty thousand slates, were despatched weekly.

The economic depression of the 1890s, and introduction of galvanised iron, resulted in a rapid decline of the industry which continued until after the First World War. Shortages of building materials at this time caused a renewal of activity at Martin's, and particularly at the larger Bangor Quarry where new plant was commissioned in 1921.

The depression years of the 1930s saw the final decline of the roofing slate industry, and only minor quantities of walling and paving stone have been produced since the Second World War. An estimated twenty to thirty million shingles and sixty thousand tonnes of walling and paving stone have been produced at Willunga. An example of roofing slate from this source is Bonython Hall at the University of Adelaide.

***189** Bangor slate quarry, circa 1921. The quarry face was forty metres above the platform level, and extended twenty metres below. Note the waste dumps below the splitting sheds which account for up to ninety-five per cent of quarried stone.*

***190** A similar view of Bangor Quarry in July 1974.*

191 A 1921 photograph of a large slab of Willunga slate prior to dressing and splitting into roofing shingles.

192 Splitting shed at Bangor Quarry on opening day of the new plant in 1921. Note the slate roof, and stockpile of roofing shingles in the left foreground.

193 Hand splitting of slate shingles at Willunga in 1920. A machine has yet to be devised which can split shingles.

Mintaro

Opened in 1856, Mintaro is the oldest continuing quarrying operation in Australia.

In the early years, most of the slate was used locally for wine vats, troughs, tanks, sills and slab walls; concrete and stainless steel are now substituted in these applications. In the late nineteenth and early twentieth centuries, the slate was widely used for street paving, particularly in Adelaide and Melbourne.

During the 1920s and 1930s, the softer near-surface material was used as electrical switchboard bases. Since the Second World War, random pieces formerly discarded as waste have been used as crazy paving, particularly in domestic gardens. Slate floor tiles are also regaining popularity.

The most acclaimed use of Mintaro slate is in the construction of high quality billiard-table tops, which are among the best in the world.

194 Mintaro slate quarry during 1872-86. The sloping quarry floor is behind the sawing yard. Hand-powered mild steel saws fed with water and abrasive sand are being used to cut slabs to size.

195 Mechanised dressing plant at Mintaro Quarry in 1913.

196 Mintaro Quarry in 1914. Slabs split from the quarry floor were raised by a steam-powered crane and stood to season for up to one year prior to dressing. The original quarry of photograph 194, partly filled with water, is in the foreground.

197 Mintaro Quarry in 1977. Natural joints in the slate are utilised in the quarrying operations to obtain slabs.

198 A rotary-headed slate rubbing machine at Mintaro in 1977. Carborundum blocks set in the rotating head are rubbing smooth a slate slab intended as a billiard-table top.

LIMESTONE AND DOLOMITE

South Australia has large reserves of limestone and its magnesium bearing variant, dolomite. Three million tonnes are mined annually for industrial applications and a similar amount is quarried for construction purposes. The principal industrial uses are in the manufacture of cement and chemicals, as fluxes in metal smelting, as whiting, and in refractories, glass and agriculture.

Limestone

Limestone and marble have been used as dimension stone and aggregate since first settlement of the Colony. By the 1880s, the developing science of cement manufacture created interest among early builders because of a lack of natural timber, and occasioned a demand for limestone to support a local cement industry.

William Lewis established a small kiln at Brighton in 1882 and manufactured Australia's first cement, but competition from overseas products forced his operations to close in the following year. In 1892, The S.A. Portland Cement Co. Ltd established a cement plant near the original kiln, based on limestone from the adjacent Linwood Quarry, augmented by quarries near Reynella which were linked to the plant by an aerial ropeway.

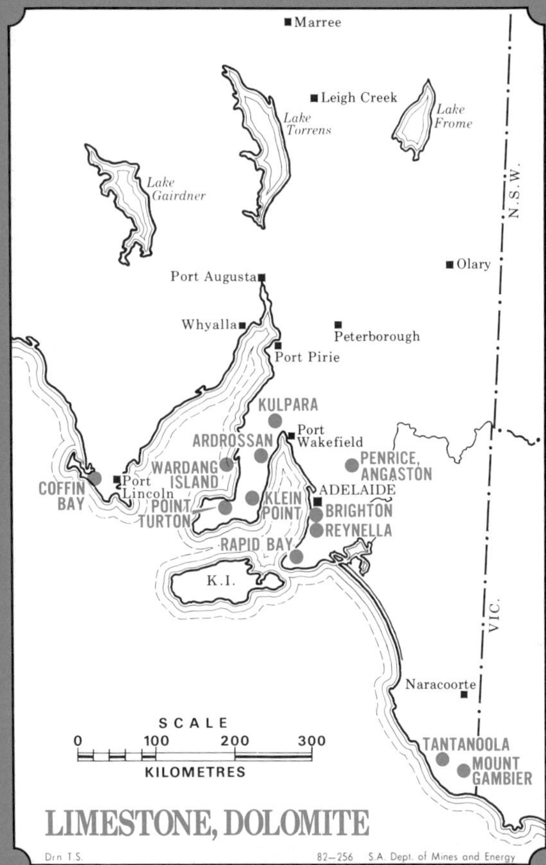

The company transferred its plant from Brighton to Angaston in 1953, where plentiful supplies of high grade marble were available from the Angaston Quarry. This quarry was opened by ICI Alkali (Aust.) Pty Ltd (ICI) in 1939 as a source of chemical grade limestone; in 1950, ICI transferred mining operations to nearby Penrice. The Penrice Quarry now supplies raw material to both the ICI chemical plant at Osborne and the cement works at Angaston; Angaston Quarry closed in 1971.

Adelaide Cement Co., formed in 1913, operated a cement plant at Birkenhead. Limestone was obtained from Stansbury and nearby smaller quarries on eastern Yorke Peninsula until 1919 when Klein Point Quarry was opened four kilometres south of Stansbury; limestone is shipped across Gulf St. Vincent to the Birkenhead plant.

The S.A. Portland Cement Co. and Adelaide Cement Co. amalgamated in 1971 to form Adelaide Brighton Cement Ltd which still operates the Birkenhead and Angaston cement plants.

The second largest industrial use of South Australian limestone is as a flux in metal smelting. Initially, the lead smelters at Port Pirie, operated by The Broken Hill Associated Smelters Pty Ltd, used limestone from Point Turton and Wardang Island, and shellgrit from near Port Pirie; limesand from Coffin Bay is now used.

Rapid Bay Quarry was opened in 1940 by The Broken Hill Proprietary Co. Ltd (BHP) to supply flux to the steelworks at Whyalla, Port Kembla and Newcastle. The quarry operated continously until 1981 when it was purchased by Adelaide Brighton Cement Ltd.

Shellgrit from Port Gawler has been the prime source of calcium carbonate for glass manufacture in South Australia, and has been supplemented by smaller quantities of shellgrit and limestone from Mount Gambier.

Whiting, which is finely milled limestone used as a filler in plastics, adhesives and putties, was initially obtained from lime-rich mud on lakes of southern Yorke Peninsula. The most important whiting grade limestone deposits have been developed at Mount Gambier, but selected white marble from Penrice, shellgrit from Gawler, and limesand from Coffin Bay have also been used.

The principal source of agricultural limestone since 1939 has been shellgrit deposits along the eastern shore of Gulf St. Vincent between Port Adelaide and Port Wakefield, supplemented by limestone from Mount Gambier.

Burnt lime, for use in agriculture and the construction and chemical industries, was produced from surface limestone or calcrete which was otherwise a nuisance to farmers in the ploughed paddocks of Yorke Peninsula, Eyre Peninsula, and the Murray Bridge-Tailem Bend area. The last lime kiln to be based on these materials was at Kadina; it closed in 1970. Since then all lime has been manufactured by Adelaide Brighton Cement Ltd, and is now produced at the Angaston cement plant from marble.

199 Loading raw-material bricks into bottle kilns at the Brighton plant of The S.A. Portland Cement Co. Ltd in 1896. The bricks were fired to produce clinker, then finely ground and aerated to form cement.

200 The Brighton plant was destroyed by a major fire in November 1909. The fire had a strong hold before fire fighting appliances were brought to bear, and then the water pressure was found to be frustratingly low.. The plant was rebuilt, and in full operation eight months later.

201 The S.A. Portland Cement Co. Ltd Reynella Quarry in 1928. This quarry, opened together with the nearby Hallett Cove Quarry at about 1920, supplied limestone to the company's Brighton cement works. Broken stone was carted to the overhead ropeway loading station in the background. Each skip on the ropeway carried about a half tonne of rock six kilometres north to the cement works.

202 The Brighton cement plant, adjacent to Linwood Quarry, was established in 1892; it used limestone from Linwood, Reynella and Hallett Cove quarries, and shale from Linwood.
This 1930 photograph shows the northern terminal of the aerial ropeway near a stockpile of broken limestone to the right; another ropeway carried cement from the plant to the rail station nearby. The plant was closed in 1953 when the company transferred operations to Angaston. Linwood Quarry now produces construction and road aggregate.

203 Reynella to Brighton aerial ropeway in 1928. This was used from about 1925 until the Reynella and Hallett Cove quarries closed in 1953.

204 Cement clinker mill being hauled to the Brighton plant in 1930. The mill produced a finer-grained product than previously obtainable, and cement quality was improved.

205 Angaston marble quarry, operated by ICI, in 1941. The marble was railed to Osborne for chemical manufacture.

206 Haulage house and gantry for transferring marble to the railway.

207 Adelaide Brighton Cement Ltd works and Angaston Quarry in the 1970s. The plant, which began production in 1953, utilises limestone from Penrice, shale from Truro, Moculta and Birdwood, gypsum from Blanchetown, and sand from Nuriootpa.

208 Blasting a face at Penrice Quarry in 1946.

209 Penrice Quarry in 1981. Nearly seventeen million tonnes of high grade marble have been quarried since 1950; nine million tonnes have been railed to the Osborne chemical plant of ICI, and five million tonnes have been trucked to the nearby cement works. Smaller amounts have been used as road aggregate, whiting, smelter flux, and in the manufacture of burnt lime.

210 Klein Point limestone quarry, crushers and jetty loading facilities in 1923. Adelaide Cement Co., now Adelaide Brighton Cement Ltd, has operated the quarry since 1919 to produce limestone for cement manufacture at Birkenhead.

211 Quarry operations at Klein Point in 1923. Ore trucks, hand loaded with limestone, are being winched up an incline to the crushers.

212 Loading blast holes with explosives at Klein Point in 1928.

213 Power shovel loading locomotive-drawn ore trucks at Klein Point in 1946. The single face quarrying method was still employed at this time.

214 Klein Point Quarry in 1977. Twelve million tonnes of limestone have been mined since 1925 and shipped across Gulf St. Vincent to Birkenhead for cement manufacture. At the jetty is MV 'Accolade', the Adelaide Brighton Cement Ltd limestone carrier which makes six return trips to Adelaide each week. The jetty gantry is stationary and the ship is winched past the loading chute to distribute the load. The company has commissioned the larger capacity MV 'Accolade II', which replaced its predecessor during 1982.

215 BHP Point Turton limestone quarry in 1915. The quarry was opened in 1906 as a flux producer for smelters at Port Pirie and in the eastern States. Over one hundred thousand tonnes were produced before operations ceased in 1919. The Point Turton caravan park is now on this site.

216 Wardang Island limestone quarry, operated by BHP, circa 1918. Limestone was first mined here in 1910, and to closure in 1968 over one million tonnes had been produced.

217 Quarrying limestone at Rapid Bay in 1951. The rock was used as smelter flux at Whyalla, Port Kembla and Newcastle from 1940 until the quarry was sold by BHP to Adelaide Brighton Cement Ltd in 1981; fifteen million tonnes have been produced.

218 *Rapid Bay Quarry in 1975. Two or three ships were loaded each month with up to eight thousand tonnes of limestone. Rapid Bay township is at lower left.*

219 Lime kilns at Kulpara in 1957. For many years, South Australian lime production was based on calcrete gathered from ploughed fields. The majority of kilns for burning the rock were on Yorke Peninsula, but others were at Tailem Bend, Murray Bridge, Gawler, Port Lincoln and Naracoorte. The principal producer was Wool Bay Lime Co. which operated these kilns at Kulpara.

Hand-sized limestone was loaded through the side and top of the kilns and fired with local wood.

Dolomite

The largest industrial dolomite quarry in the State, that at Ardrossan, was opened in 1950 by BHP to supply flux to the steelworks at Whyalla, Port Kembla and Newcastle, and the aluminium refinery at Bell Bay (Tasmania). High purity dolomite from Ardrossan is manufactured into refractory bricks at Port Kembla.

Dolomite has been quarried near Tantanoola since 1961 for plate glass manufacture in Melbourne. Small amounts have been mined near Nuriootpa and Mount Gambier for agricultural top-dressings.

220 Dolomite crushing plant at Tantanoola in 1977.

221 BHP Ardrossan dolomite quarry in 1979. The quarry has produced nine million tonnes since 1950; most has been used as flux.

GRAPHITE

Disseminated flake graphite occurs at several localities on southern and central Eyre Peninsula, but mining has been restricted to the Uley and Koppio areas. Mining commenced in 1866, with greatest production recorded during the two world wars. High production costs and restoration of overseas supplies after the Second World War forced closure of mining operations.

Total production amounted to eight hundred tonnes of refined graphite. This was used in crucibles and mould facings by the foundry industry, and in dry cell batteries, lubricants, carbon brushes and points.

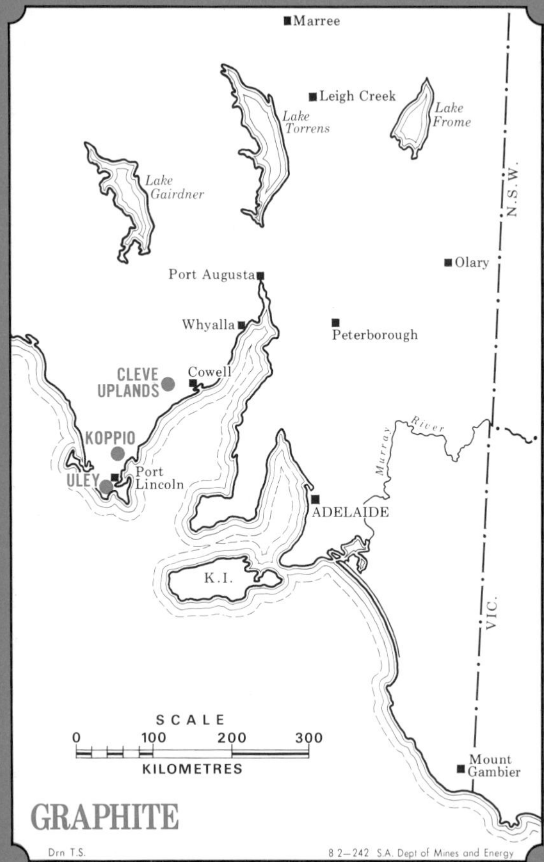

222 Two of the eight shafts at Uley graphite mine in the 1940s.

223 Uley graphite mill and refinery in November 1928. The plant was erected in 1928 near the Uley Mine, twenty kilometres southwest of Port Lincoln. Refined graphite was recovered by flotation.

SALT

Before 1966, South Australia produced eighty per cent of Australian salt. The bulk of production now comes from Western Australia.

Small quantities were scraped from the surfaces of lagoons near Kingscote on Kangaroo Island as early as 1814 for domestic purposes, meat preservation, and tanning of animal hides. Regular harvesting from salt lakes commenced in 1891 on southern Yorke Peninsula.

To supplement this natural salt production, seawater evaporating ponds were constructed at Port Augusta and Price during the First World War. In 1940, production of salt for the alkali industry commenced near Adelaide from fields at Dry Creek, and these are now the largest producers in South Australia. Salt is also recovered from Lake Macdonnell by evaporation of brines pumped from nearby gypsum workings.

Salt derived by solar evaporation of seawater now accounts for most of the State's production, and only Lake Macdonnell and Lake Bumbunga yield significant tonnages of natural salt.

The chemical industry consumes most of the harvest for the manufacture of sodium salts and chlorine.

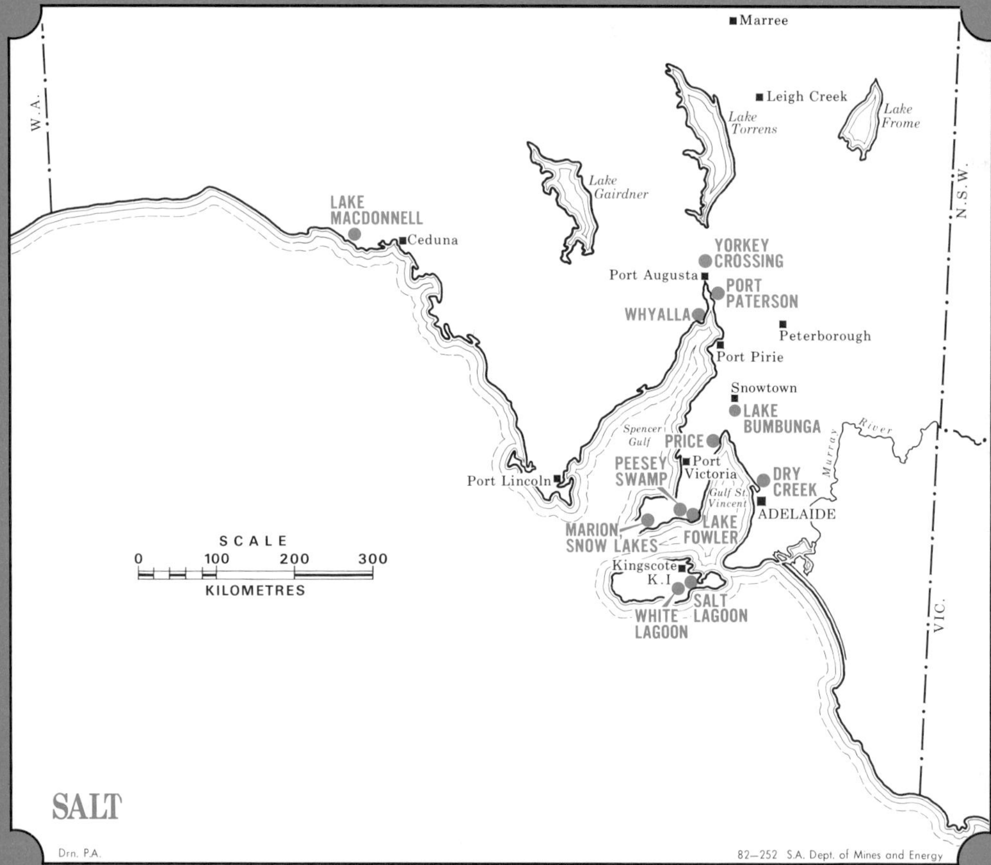

Yorke Peninsula

Natural salt crusts form readily in this area owing to high winter rainfall combined with hot summers for evaporation.

Lake Fowler is the largest saline lake, and crude salt was first harvested here over one hundred years ago. Regular large scale seasonal production dates from 1891.

Peesey Swamp, Marion Lake, and Snow Lake were also worked in the nineteenth century. Since the early 1930s, Australian Salt Co. Ltd harvested and trucked salt to Edithburgh for refining and shipment; production ceased in 1965.

At present, only four small lakes near Warooka and Port Victoria produce salt.

***224** Harvesting at Salt Lagoon, near Yorketown, in 1916. Shovels were used to scrape the salt crust into heaps, which were left for a time to allow the more soluble magnesium salts to drain away. The remant, relatively pure common salt was bagged for market.*

***225** Motorised salt harvesting on Lake Fowler in the 1930s. The vehicle at left is scraping salt towards a bucket elevator which discharges into the rear-tipping dump trucks.*

***226** Salt scraping at Peesey Swamp in 1957. The salt is being shovelled into side-tipping railcars for cartage to the lake edge for stockpiling and drainage. The tracks are made in short sections to facilitate movement across the lake during harvesting.*

Lake Bumbunga

Saline lakes in the Lochiel-Snowtown area have been harvested since the early 1900s. Lake Bumbunga is now the only significant producer. Australian Salt Co. Ltd began operations in 1913 from evaporating pans and crystallisers constructed on the southern portion of the lake. At first salt was gathered by shovel and later by horse-drawn scoops; these were replaced by motorised harvesters and dump trucks.

227 Loading salt into railcars in the 1920s. Horse-drawn scoops are scraping salt over a grid on top of the mobile ramp to load the railcars underneath.

228 Employees at Lake Bumbunga holding salt-encrusted tree roots and stumps in 1920.

229 Salt harvesting in the 1920s. The salt crust is being broken by a horse-drawn scarifier in the foreground, and collected by a scoop in the background. The scarifier is still used, but the horses have been replaced by a tractor.

230 Mechanical salt harvesting in the 1930s. The bucket elevator, which lifts salt into the partly hidden dump truck, is in front of the operator. This method of harvesting is still used on Lake Bumbunga.

231 Salt stockpiles on the lake shore in 1938. Harvested salt is being tipped onto the conveyor apron for stockpiling prior to refining.

232 Northeasterly view of Lake Bumbunga in 1981; the small town of Lochiel is at bottom centre. The refinery and stockpiles are on the western lake shore opposite the large, square evaporating pans and crystallisers.

Kangaroo Island

South Australia's first salt harvest was recovered in 1814 from White Lagoon. The nearby Salt Lagoon deposits were worked by the Globe Salt Co. from 1897, and production continued under different companies until 1961. The output was shipped to Port Adelaide for refining until a refinery was built on the shore of the lagoon in the early 1920s, and a light railway constructed to Muston on American River to facilitate shipment.

Several smaller deposits were also worked near Kingscote and Salt Lagoon but production has now ceased.

233 ***Steam locomotive on the Muston Railway, used to haul bagged salt from the Salt Lagoon refinery to American River on Kangaroo Island, in the early 1920s.***

234 ***This engine, with a converted tractor motor, replaced the earlier steam locomotive on Muston Railway. The engine was capable of pulling three railcars, each carrying five tonnes of bagged salt.***

235 ***Harvested salt stockpile at Salt Lagoon.***

236 ***Salt refinery on the shore of Salt Lagoon; date of the photograph is unknown.***

Solar Evaporation of Seawater

The northern shores of Gulf St. Vincent and Spencer Gulf are ideal sites for evaporation ponds, where extensive flat areas, underlain by impervious mud to restrict water loss, are available. Evaporation rates are high and seawater in the gulfs has an above average salt content.

The first solar ponds were constructed in 1915 at Yorkey Crossing, north of Port Augusta, and were used until 1930. A smaller operation near Port Paterson lasted from 1951 to 1963.

The Dry Creek fields operated by ICI Australia Operations Pty Ltd (ICI) are the largest salt producers in South Australia. Annual output is more than a half million tonnes, most of which is redissolved and pumped as a brine under the Port River to the Osborne plant for chemical manufacture.

Evaporation ponds constructed at Price in 1919, and since extended, are operated by Ocean Salt Pty Ltd.

The Broken Hill Proprietary Co. Ltd (BHP) produce solar salt from ponds constructed north of Whyalla.

237 The Ocean Salt Pty Ltd factory, stacker, and evaporation ponds at Yorkey Crossing, north of Port Augusta, in 1923. Operations were closed in 1930 by encroaching sand dunes which partly covered the ponds.

238 Salt harvesting from evaporation ponds near Price in 1933. Salt is being shovelled onto the moveable conveyor for stacking.

239 *Official opening of the ICI Dry Creek saltfields in 1936. Seawater is being pumped into the first evaporating pond.*

191

240 View across the Dry Creek saltfields in 1952. Torrens and Garden islands, Port River, and Le Fevre Peninsula are in the background.

241 View across the Dry Creek saltfields in 1982. The fields cover about four thousand hectares; crystallisers, where the salt is finally precipitated, are at left centre. Forty-one million tonnes of seawater were pumped through the ponds in 1979.

242 Mechanical harvesting at Dry Creek in 1979. Spiral screws guide salt towards bucket elevators which deposit their load onto a moveable conveyor for stacking. Prior to stripping of the salt, furrows are dug to drain excess brine from below the salt crust. Harvesting is carried out between March and June each year.

243 Salt being heaped into stockpiles by a mobile conveyor-stacker. The stockpile is leached for a few months to remove most of the magnesium salts. Jets of fresh water redissolve the remaining salt to produce a concentrated brine which is piped under the Port River to the Osborne ICI chemical plant.

244 Osborne ICI chemical plant on the western bank of the Port River in 1981. The plant is linked by a ten kilometre brine pipeline to the saltfields in the background. Torrens Island Power Station is in the middle distance.

245 Southerly view of the BHP Whyalla evaporating ponds in 1973. Feedwater from cooling circuits in the steelworks, two kilometres south, is channelled to the ponds. Salt is harvested from crystallisers in the right centre of view.

OIL AND NATURAL GAS

Early petroleum exploration in South Australia was concentrated along the coast from Port Wakefield to Mount Gambier, and Australia's first well in search of petroleum was drilled to a depth of over two hundred and eighty metres at Salt Creek near The Coorong between 1881 and 1883; no traces of oil or gas were found.

One of the early wells, drilled east of Robe in 1916, reached a total depth of nearly fourteen hundred metres, and was the deepest in the State until 1959. Another well collared near Kingston in 1925 took five years to complete. Wells were also drilled on Kangaroo Island, Yorke Peninsula, and Eyre Peninsula.

Petroleum exploration received a major boost with the formation of South Australia and Northern Territory Oil Search (Santos) Ltd in 1954. This company and its partners were responsible for the discovery and development of the Cooper Basin natural gas field. Although the first eight wells were barren, commercial quantities of gas were found at Gidgealpa in 1963. Further drilling defined the extent of the field, and in 1968 the Moomba gas treatment plant was established to supply gas by pipeline to Adelaide. Another pipeline was connected to Sydney, and supplies commenced in 1976.

In 1978, natural gas from Moomba generated almost seventy-five per cent of the South Australian electricity output, and supplied over a third of its total energy requirements.

Oil was discovered at Tirrawarra in the Cooper Basin during 1970, and a number of other oil fields have since been proved. The oil will be pumped, along with gas condensate and liquefied petroleum gas, to the Stony Point fractionation plant on Spencer Gulf for separation and refining prior to sale to Japanese and local consumers.

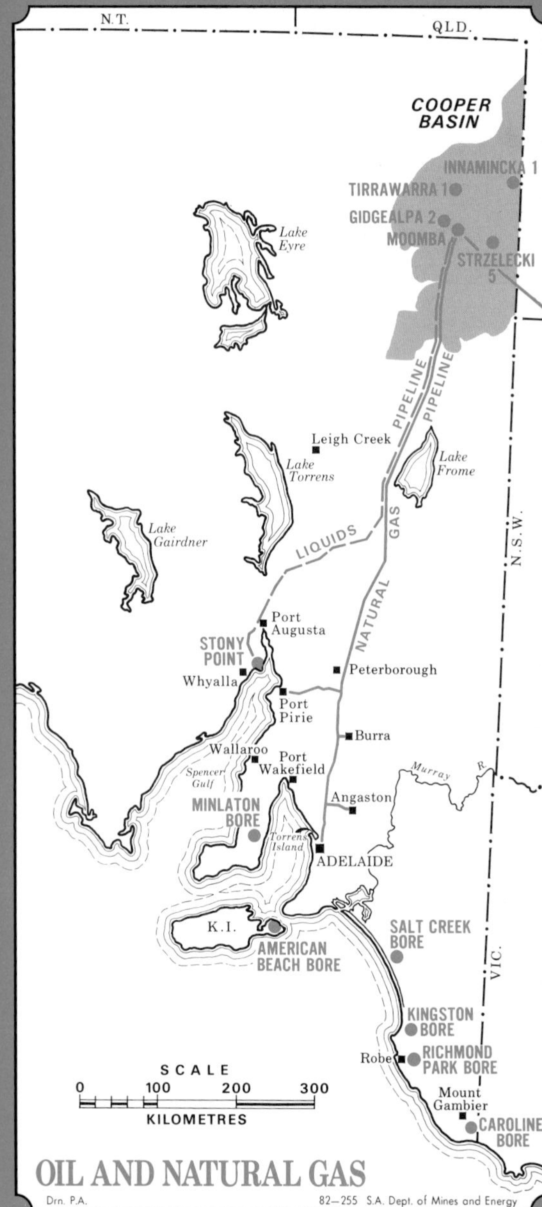

246 *This exploration well at Richmond Park near Robe was completed in 1916. Traces of tar and wax were found in the drill cuttings. With a total depth of nearly fourteen hundred metres, it remained the deepest well in South Australia until 1959 when it was surpassed by Innamincka No. 1 in the Cooper Basin.*

247 *The American Beach petroleum exploration well being drilled by a steam powered rig on Kangaroo Island in 1921. Total depth of the well was nearly three hundred metres, and a very small gas flow was recorded.*

248 *At the time that South Australia's early petroleum wells were being drilled, petrol refined from imported crude oil was carried to Adelaide's few distributing stations in tankers such as the one in this 1917 photograph.*

249 Collecting gas from a petroleum exploration well near Minlaton in 1931. The rig was testing for an oil pool thought, by a number of diviners, to exist on the property. The well reached five hundred and fifty metres, and a small gas flow of almost pure hydrogen was recorded.

250 The first petroleum exploration well in the Cooper Basin, Innamincka No. 1, in 1959. Its final depth was almost four kilometres; only trace amounts of gas were found and the well was abandoned.

251 The first commercial natural gas well in Australia was completed at Gidgealpa in 1964. The arrangement of pressure valves at the top of the well is known in the petroleum industry as a 'Christmas Tree'.

252 Moomba natural gas treatment plant in 1981. Gas from satellite stations on fields in the Cooper Basin is piped to Moomba through a system of gathering and trunk lines. The plant removes carbon dioxide, hydrogen sulphide, hydrocarbons and water to produce market gas.

253 A 1981 aerial view of the Torrens Island Power Station. In that year the station consumed nearly two-thirds of the gas carried by the Moomba-Adelaide pipeline, and provided three quarters of the State's electricity.

254 A pneumatic breaker clearing the Moomba-Stony Point liquids pipeline trench of hard rock.

256 Curved sections of the pipeline being assembled.

255 The completed trench, and pipeline ready for welding and laying.

257 *Stony Point, site of the Cooper Basin liquids fractionation plant, on the shore of Spencer Gulf.*

258 *The offshore oil drilling platform 'Ocean Digger' near Robe in 1967. The platform, constructed at the Whyalla shipyards in 1967, has drilled four wells in South Australian waters.*

203

260 In 1967, a petroleum exploration well at Caroline near Mount Gambier, completed at a depth of more than three kilometres, tapped a reservoir of carbon dioxide. Commercial production began in the following year, and output is now approximately two tonnes of purified carbon dioxide per hour for use in carbonated beverages, fire extinguishes, and for medical applications.

259 Petroleum exploration drilling rig 'Danwood Ice' at site of Mercury 1 well off the west coast of Eyre Peninusla in 1981.

COAL

Early traces of coal in South Australia were located during water-well sinking near Montefiore Hill, but the find failed to create interest. In 1885, water boring operations in the bed of Lake Ifould, near Pidinga, also brought coal chips to the surface, but remoteness of the area precluded further exploration.

Coal was disclosed during excavations for a railway dam near Copley in 1888. Thicker seams were outlined by drilling in a separate basin at Leigh Creek, ten kilometres to the north, and a small colliery operated for two periods until 1908. Continuous production of coal commenced at Leigh Creek in 1943.

Brown coal was found at Noarlunga in 1905 and Hope Valley in 1911. Similar coal was located during water boring operations at Clinton and Inkerman, and on the western margin of the Murray Basin at Bower in 1902, Moorlands in 1910, and Anna in 1920. During the mid 1970s, more extensive brown coal deposits were discovered at Kingston and Sedan; these are currently being investigated for possible development.

A brown coal deposit was located at Lake Phillipson, southwest of Coober Pedy, while boring for water in 1905; a very large deposit was outlined in this area during the 1970 and is still being tested.

Coal has been mined intermittently since 1923 from a shaft near Lock on Eyre Peninsula to provide domestic fuel. Exploration in the 1970s outlined the extent of the seams.

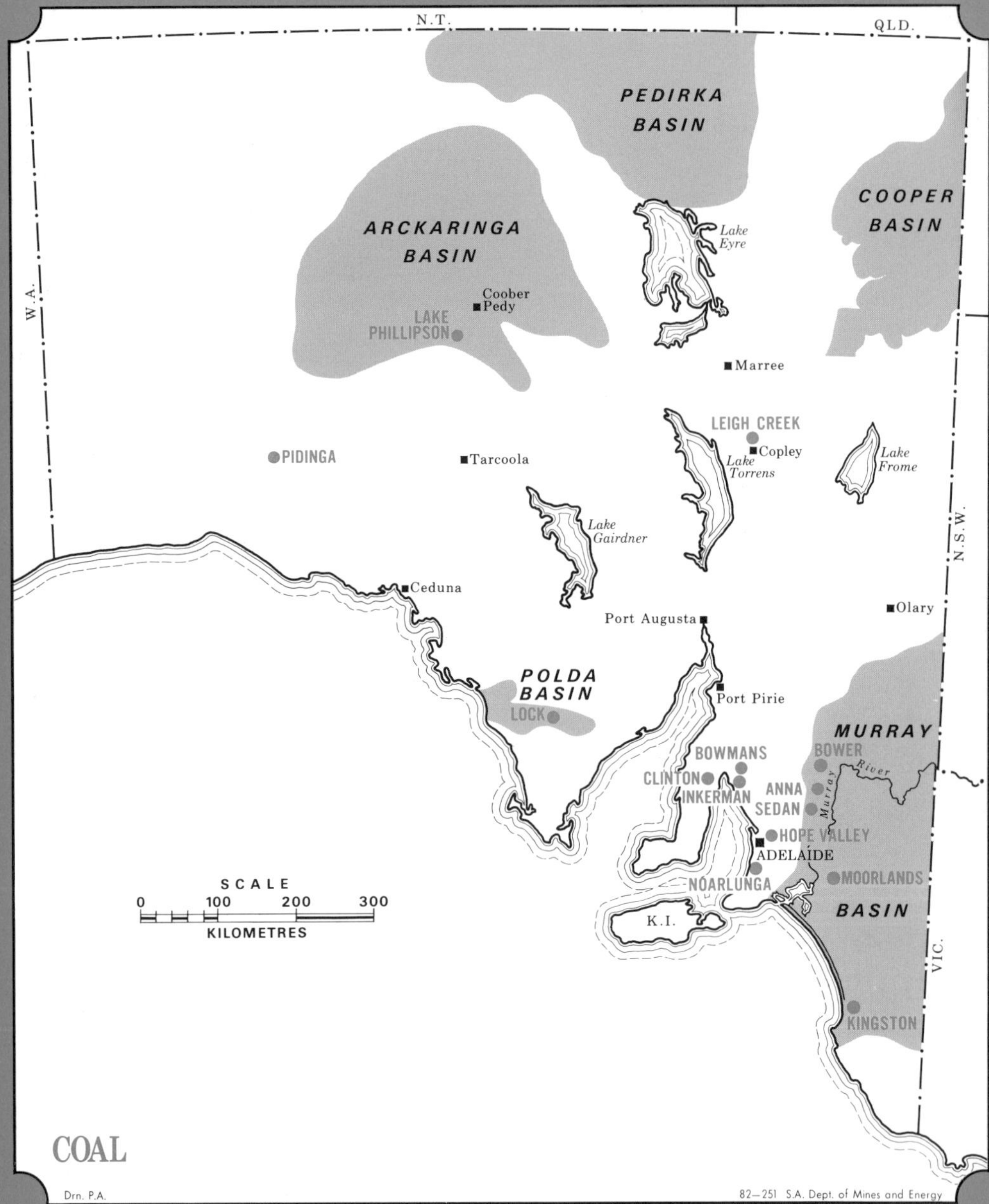

Petroleum drilling in the Cooper and Pedirka Basins has intersected huge reserves of high quality black coal at depths of over one and a half kilometres, too deep for profitable recovery by current technology.

Leigh Creek

Shortly after the coal discovery near Copley in 1888, the first Government Geologist, H. Y. L. Brown, mapped the extent of the coal measures in four separate basins and predicted that the field would prove to be of major proportions.

A shaft was sunk by the Leigh Creek Coal Mining Co. in 1892, and coal was processed in an experimental briquetting plant. The coal proved unsuitable for this purpose and the mine closed in 1894.

In 1906, the lease was transferred to the Tasmanian Copper Co., which raised over twelve thousand tonnes of coal by 1908 when the lease expired. This was used for domestic purposes, although some was mixed with better quality Newcastle coal for fuelling steam locomotives.

During 1910-18, exploration holes were drilled to define the size of the field, but interest waned until 1941 when extensive drilling and shaft sinking were undertaken.

Open cut mining began in 1943 under the management of the Engineering and Water Supply Department, which was responsible for water supply to the mine and town of Leigh Creek. Coal was sold as an industrial and domestic fuel, and for use in the modified boilers of Osborne Power Station.

The Electricity Trust of South Australia (ETSA) took control of operations in 1948 and currently use the entire production, approximating two million tonnes annually, for their power stations at Port Augusta.

In 1974, the decision was made to deepen the open cuts to mine previously inaccessible reserves. The township of Leigh Creek was moved to a new site fifteen kilometres south of the old town, and was officially reopened as Leigh Creek South on 31 October 1981 by the Premier, Dr David Tonkin.

At the projected rate of mining, Leigh Creek coal reserves will last for at least thirty years.

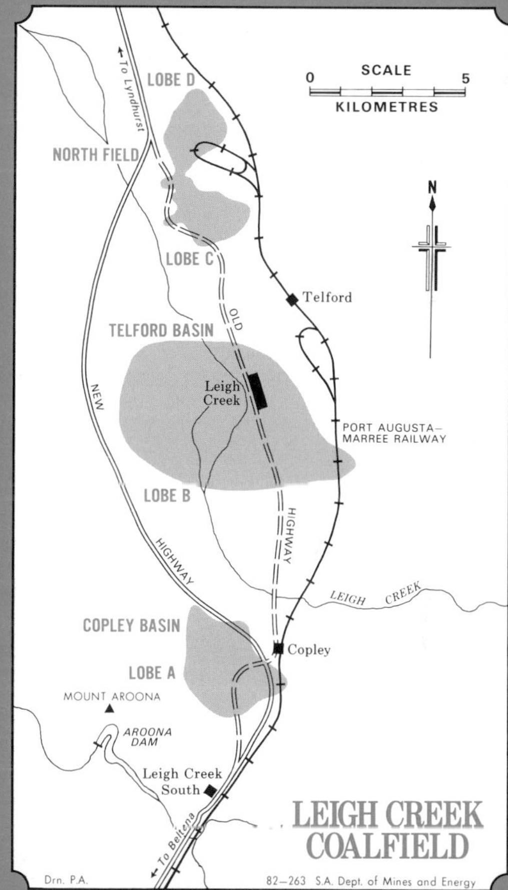

261 The abandoned coal shaft at Leigh Creek in late 1894. The shaft was sunk in 1892 and two hundred tonnes were raised and processed in a briquetting machine in the right foreground. The coal was found to be unsuitable for this purpose, and operations ceased in 1894.

262 Telford Cut, the first on Leigh Creek Coalfield, in 1945. A small dragline for loading coal into dump trucks is on the pit floor. The main coal seam was mined to a depth of thirty metres. Operations were suspended in the mid 1960s and transferred to North Field, but mining recommenced in the Telford Basin during 1976.

263 Lobe D at North Field in the early 1960s. Production commenced in 1956 and continued until the coal seams were exhausted in 1976. Coal was mined from two seams; the lower seam is exposed in the bottom of the cut and is separated by shale, exposed as a bench at left, from the upper seam. A dragline at right is removing overburden from the upper seam, while coal is being won from the lower seam in the right foreground.

264 Coal being mined from the upper seam of Lobe D at North Field in 1960. The dragline in the background is removing overburden from the darker coal seam which has been completely exposed in the foreground. This is being mined by an electric shovel and front-end loader partly hidden by the seam.

265 Drilling blast holes in hard overburden at Leigh Creek.

266 Loading blast holes with explosives in one of the large open cuts at Leigh Creek.

267 A load of over one hundred and fifty tonnes of overburden being hauled from an open cut.

268 Coal crushers at Telford Basin in 1980. These were installed in 1976 adjacent to a spur of the Leigh Creek-Port Augusta railway. Conveyors in the foreground are carrying coal to the secondary crushers and loading facilities.

269 Crushers, loading facilities, and coal train at Telford in 1980. The bins, at left, load eight railcars at a time. A fully laden train carries eighty railcars with four thousand tonnes of coal to the Port Augusta Power Station.

270 The Port Augusta coal-burning power stations in 1982. Thomas Playford A and B stations, commissioned in 1954 and 1960, are in the foreground, and the partly constructed Northern Power Station is behind. With commissioning of the new station in 1984-85, electricity generation at Port Augusta will approximate one third of the State's capacity.

271 Osborne Power Station in 1982. From 1944, coal was railed to Osborne to fuel the modified boilers which previously burned higher grade Newcastle black coal. Osborne was converted to oil in 1972, and has been classified as a standby station since the larger Torrens Island natural-gas-fired power station was commissioned in 1967. To the left of the power station is the lighter coloured chemical plant of ICI Australia Operations Pty Ltd, which utilises salt from nearby Dry Creek as a raw material.

272 Southwesterly view from Mount Aroona over the proposed Aroona dam site in 1948. At this time, water was supplied to the coalfield and town from local dams and bores. In addition, water was pumped from Sliding Rock copper mine, nearly fifty kilometres to the southeast.

273 Aroona Dam at near capacity in 1967. The dam has supplied Leigh Creek Coalfield and township with water since November 1954. The dam wall is nearly twenty metres high, but silting in recent years has reduced the reservoir's capacity.

274 Aerial view southwesterly of Leigh Creek in 1957. The Hawker-Lyndhurst road and railway are on the far side of the town, and Leigh Creek meanders into the distance towards the Copley Range which contains Aroona Dam. The town has now been shifted south to allow mining of coal under this site.

275 Easterly view of the new township of Leigh Creek South in 1982. The main road from Hawker is in the centre of view.

Coal near Adelaide

As coal to supply the Adelaide market was in short supply at the turn of the century, great efforts were made to find a local source. A deposit was worked for a short time at Montefiore Hill in 1894, but reserves were small and the quality poor.

In 1905, the Noarlunga Coal Co. extracted forty tonnes of brown coal from two shafts on the banks of the River Onkaparinga near Old Noarlunga; as with many South Australian deposits, the seam was discovered during water boring operations. Drilling in 1922 showed that the deposit was too small to justify development.

Water bores near Hope Valley intersected brown coal in 1911, and two shafts were sunk to obtain bulk samples. Total production was apparently very small.

The Wakefield coal deposits were discovered at the beginning of this century, and government drilling at Clinton and Inkerman partially outlined their extent. During 1975-76, the Department of Mines and Energy drilled near Port Wakefield for ETSA and defined an extensive deposit of brown coal. A bulk sample was taken in 1980 from a circular test pit at Bowmans for trial burning to determine suitability for power generation.

276 Exploring for coal at Moorlands, southeast of Murray Bridge, in 1922. The drilling rig is powered by a kerosene engine.

277 Government coal boring at Hope Valley in 1920. The horse-drawn kerosene engine provided power to the cable tool drill. Hope Valley Reservoir is in the left background.

278 Drilling with a cable tool rig at the Clinton Coalfield in 1922. The coal being tested extends under Gulf St. Vincent to Bowmans, east of Port Wakefield.

279 Northerly view of Bowmans trial coal pit in July 1980, midway through the excavation which was completed in November 1980. The coal was sent to West Germany and USA for test firing to determine suitability as a power station fuel. Many pumps were employed to control flow of groundwater and to maintain stability in the pit walls during excavation. Dewatering pipes encircle the pit; further pumps were installed in the pit as depth was increased to the limit of sixty metres.

280 The trial pit in November 1980, four days after removal of the dewatering pumps. Natural groundwater flow quickly filled the pit, causing unconsolidated sand in the walls to collapse.

281 The trial pit in January 1981, two months after mining was completed and the dewatering pumps removed. Groundwater has now reached its original hydrostatic level.

282 Brown coal exposed in the bottom of the trial pit. The seam is about ten metres thick; the top is above the arm of the backhoe, but the bottom was not exposed during excavation.

IRON

Middleback Ranges

The iron ore deposits in the Middleback Ranges west of Whyalla have been important in the development of mining and manufacturing in South Australia. The year of their discovery is unknown, but in 1840 explorer Edward John Eyre, who camped at what became Iron Baron, described the rocks as 'principally iron-stone formation'. The name Iron Knob first appeared on pastoral lease maps of 1854.

The large quantity and high grade of iron ore was recognised by the late 1890s, and The Broken Hill Proprietary Co. Ltd (BHP) pegged the first claim in 1897. Mining commenced in 1900 and ore was carted by bullock wagon to Port Augusta then by rail to the Port Pirie lead smelters for use as a flux. Some pig iron was produced experimentally at Port Pirie in 1907.

A tramway from Iron Knob to Hummock Hill (renamed Whyalla in 1914) was completed in 1901. Shipping wharves were opened in 1903 and these allowed larger scale mining operations to proceed, with barges carrying ore directly to Port Pirie.

Dwindling ore reserves at the Broken Hill silver-lead-zinc mines operated by BHP caused the company to diversify, and in 1912 steelworks were established at Newcastle, NSW, using South Australian iron ore.

In 1920, iron ore was discovered at Iron Baron, Iron Prince and Iron Queen, and in 1934 at Iron Duchess, Iron Duke and Iron Knight. The ironstone outcrops have now been traced over a distance of sixty kilometres.

Nearly three million tonnes of ore are mined annually from Iron Knob, Iron Monarch, Iron Baron, Iron Prince and Iron Queen for the steelworks at Whyalla; a small amount is shipped to the eastern States and overseas.

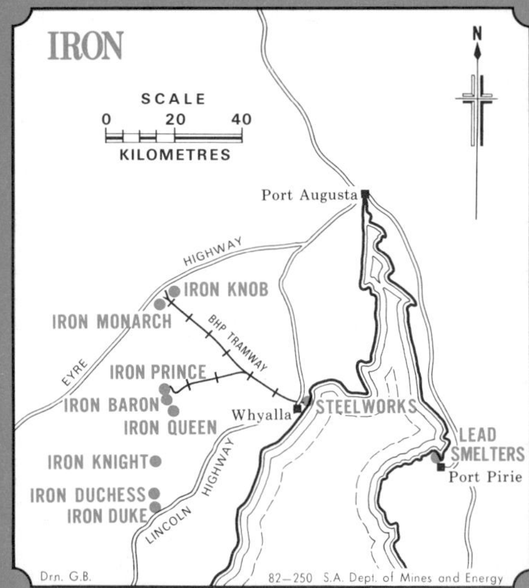

283 Southerly view of Iron Knob hill in 1900. A prospector's hut with corrugated iron chimney is in the foreground.

284 Iron ore train travelling from Iron Knob to Whyalla in 1933. The steam locomotives have been replaced by diesel.

285 Iron Prince Quarry in 1940. Miners scaled the face with ropes to drill blast holes, a job which proved both dangerous and time consuming. Broken rock was hand loaded into horse-drawn trucks on temporary rails and hauled to the crushing station.

287 Overburden being loaded by electric shovel at Iron Monarch in 1980.

286 Ore being loaded by steam shovel at Iron Baron in 1940.

288 Northeasterly view of Iron Monarch Quarry in 1954; crushing facilities in the foreground have since been dismantled. The high working faces were characteristic of this period.

289 Northerly view of Iron Monarch Quarry in 1976. Modern crushing facilities are at bottom left, Iron Knob Quarry is at the upper right, and Iron Knob township is to the extreme right. The Eyre Highway linking Port Augusta and Perth is in the background.

290 Blasting at Iron Monarch in 1966. A sheet-like spray of rock has emerged from one blast hole near the centre which intersected a fracture.

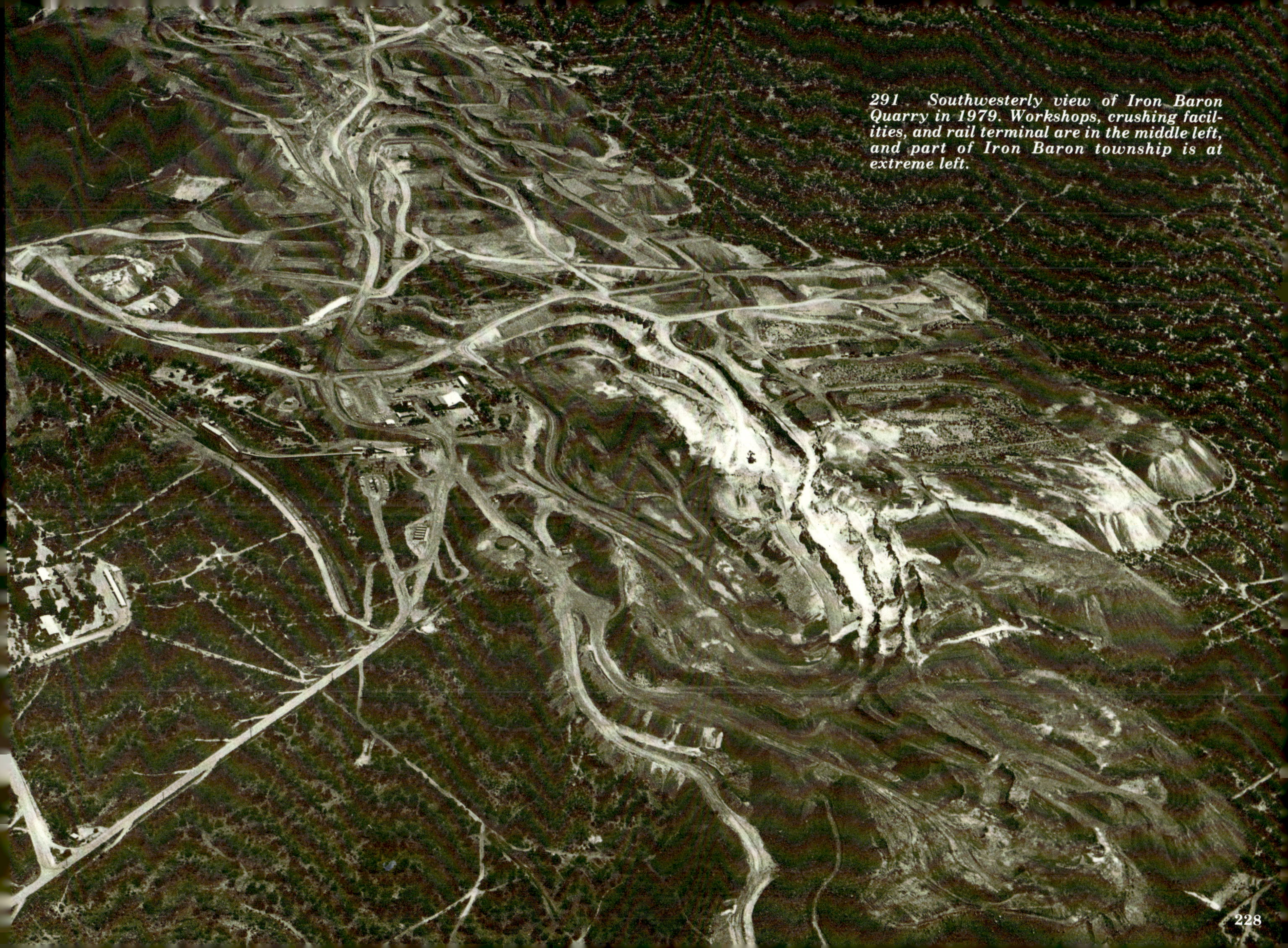

291. Southwesterly view of Iron Baron Quarry in 1979. Workshops, crushing facilities, and rail terminal are in the middle left, and part of Iron Baron township is at extreme left.

Whyalla Steelworks

In 1937, the proposal to build a blast furnace at Whyalla was announced, and the Federal Government approached BHP to build a shipyard adjacent to the proposed site. Construction began in 1938, and the first blast furnace was blown-in on 26 May 1941. Four corvettes for the Royal Australian Navy, the first appropriately named HMAS Whyalla, were launched from the new shipyards in that year.

An electric furnace and forge shop, constructed near the blast furnace to produce munitions during the Second World War, were not put into operation during hostilities; they were purchased from the government by BHP in 1945. The first steel from this furnace was poured in 1949, and a foundry for casting special steel alloys was opened in 1952. A fully integrated modern steelworks with rolling mills was commissioned in 1964, and a pellet plant to convert friable iron ore into lump form suitable as a blast furnace feed was commissioned in 1968.

In 1978, BHP announced that shipbuilding at Whyalla was to cease after sixty-six ships and an offshore oil-drilling rig had been constructed since 1941. The 'Clutha Capricorn', the largest ship built in Australia, was launched from the yards in 1972.

Whyalla became Australia's major rail-rolling centre in 1979, and over five thousand people were employed at the steelworks in 1982.

292 Pellet plant and jetty facilities at Whyalla in 1979. The pellet plant processes iron ore into lumps suitable for blast furnace feed.

293 Whyalla blast furnaces in 1975; the two furnaces are right of centre. Coke ovens to supply blast furnace fuel are partly hidden behind the steam cloud, and coal stockpiles are to the left; ore and flux stockpiles are on the wharf in front of the coalyard. Shipbuilding yards, with the partly constructed 'Union Rotorua', are at lower left. Solar evaporation salt pans are in the background to the left of Spencer Gulf.

294 Pouring molten steel into ingot moulds at the Whyalla steelmaking plant. Approximately forty minutes are required to produce each batch.

295 Steel rolling mills at Whyalla in 1979. Most of the rail for Australian railways is now produced here.

296 The offshore oil drilling platform 'Ocean Digger' under construction at the Whyalla shipyards in 1967.

297 Over one million tonnes of shipping were constructed at Whyalla. The last ship to be built was the bulk ore carrier 'Iron Curtis', pictured in 1978.

MANGANESE

The South Australian manganese mine at Boolcunda Creek was worked from 1886 to 1890, firstly by open cut and later by underground methods, to produce twelve thousand tonnes of ore.

The largest manganese deposit in the State is at Pernatty Lagoon. The Australian Manganese Co. Ltd commenced mining in 1914, when one hundred tonnes of high grade ore were sold to The Broken Hill Proprietary Co. Ltd. During the period 1915 to 1924, five thousand tonnes of ore were raised and carried thirteen kilometres by donkey and camel carts to Woocalla railway siding for transport to Port Augusta. From there it was shipped to the eastern States for steelmaking, or to Great Britain for the manufacture of munitions. Some high-grade ore was used in dry-cell batteries, and low-grade material was added to fertilizers as a trace element component.

The mine was again worked during the Second World War when twenty-seven thousand tonnes were mined for steelmaking. A further eight hundred and fifty tonnes were mined from 1968 to 1972 for fertilizers. Total production from Pernatty Lagoon amounted to thirty-three thousand tonnes.

Blinman manganese mine was the third largest producer in the State. Worked from open cut and underground during 1940-51 and 1960-61, output totalled fifty-three hundred tonnes of ore. About twenty-five hundred tonnes were mined in the same periods from Muttabee, and a similar tonnage was raised by open cut and underground methods near Bendleby in the early 1940s. Smaller manganese deposits were worked near Blinman, Yanyarrie, and Ketchowla in the 1940s.

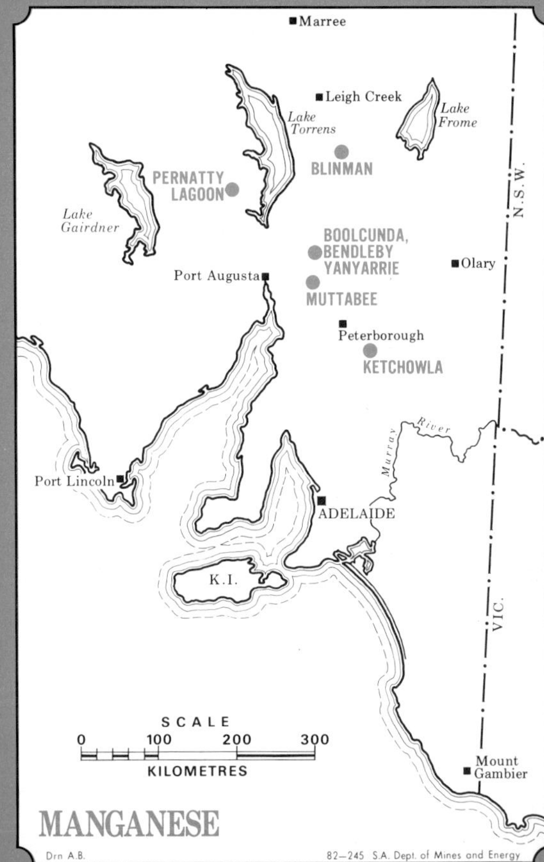

298 Main quarry of the Australian Manganese Co. Ltd at Pernatty Lagoon in 1917. The workings were in the bed of the lagoon, and excavated only to the watertable, a depth of less than two metres. The ore was mined with crowbar, pick and shovel, then hand-sorted and bagged.

299 Pernatty Lagoon manganese deposit in 1917. L. K. Ward, the Government Geologist, is at right centre facing the camera.

300 Bags of manganese ore being loaded onto a donkey cart at Pernatty Lagoon in 1917 for transport to Woocalla railway siding.

MAGNESITE

Magnesite was first mined in 1915, west of Tumby Bay. Most of the production since then has come from Robertstown, Mundallio, Copley, and Myrtle Springs. The total recorded South Australian production is twenty-eight thousand tonnes.

The largest reserves are on Balcanoona Station, one hundred kilometres east of Copley, where twenty million tonnes have been outlined. Other large deposits occur near Mount Fitton, Witchelina, and near Port Germein.

Magnesite is used in chemicals for leather tanning, welding electrodes, conversion to caustic magnesia, as a filler in rubber, in insulation fibre, stock fodder and fertilizers.

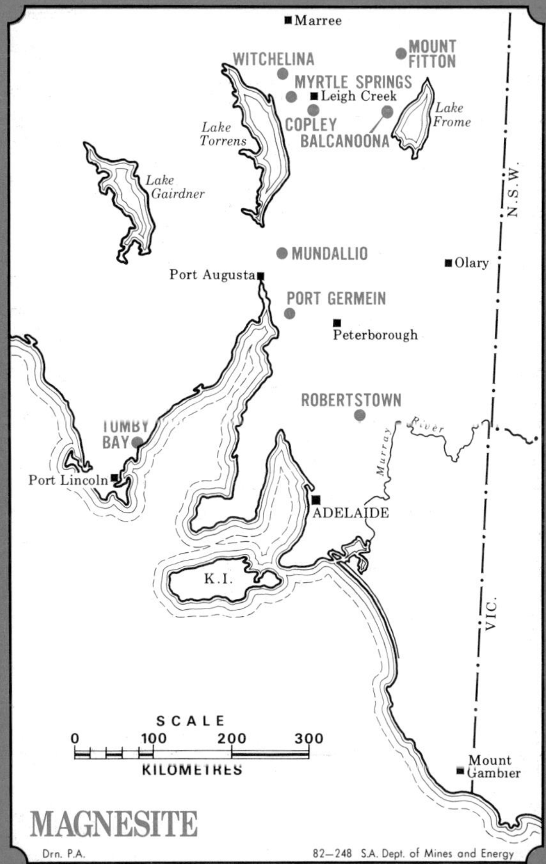

301 The abandoned Mundallio magnesite mine at about 1960.

TALC

Gumeracha was the main source of South Australian talc from 1900 until 1945 when the Mount Fitton open cut mines in the northern Flinders Ranges commenced operations. Mount Fitton is now the principal supplier of high grade talc for Australian consumption. In 1980, South Australian production approximated 10 per cent of the Australian total.

Small quantities of pure white talc have been mined near Tumby Bay, and green carving talc is currently mined at a small deposit near Cowell.

Although cosmetics are the best known talc based product, much larger quantities are used as filler in paints, rubber and vinyl. Other uses are in ceramics, as a carrier in insecticides and dusting powders, and for ornamental carvings.

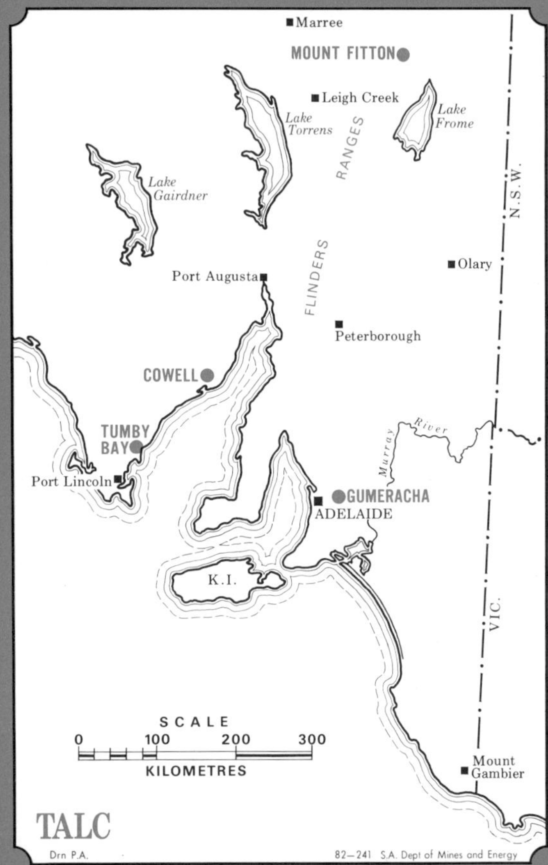

Gumeracha

Mining was originally open cut, and blocks of talc were cut into slabs for furnace linings and building stones.

By 1949, the two major mines, Dunstan's and Torrens No. 1, were worked from underground. Levels, drives and stopes followed the highly irregular talc lodes, and masses of country rock and impurities were used to backfill stopes or left as pillars. With groundwater seepage and gradual collapse of the stopes, the mines were abandoned in 1973. Since then, production has been derived from open cuts at nearby Steatite Hill and Chapman deposits, and from Tweedie Gully near Lyndoch.

Total output from the Gumeracha district is one hundred and seventy thousand tonnes, with twenty-one hundred tonnes being produced in 1980. The Tweedie Gully deposit has yielded twelve thousand tonnes.

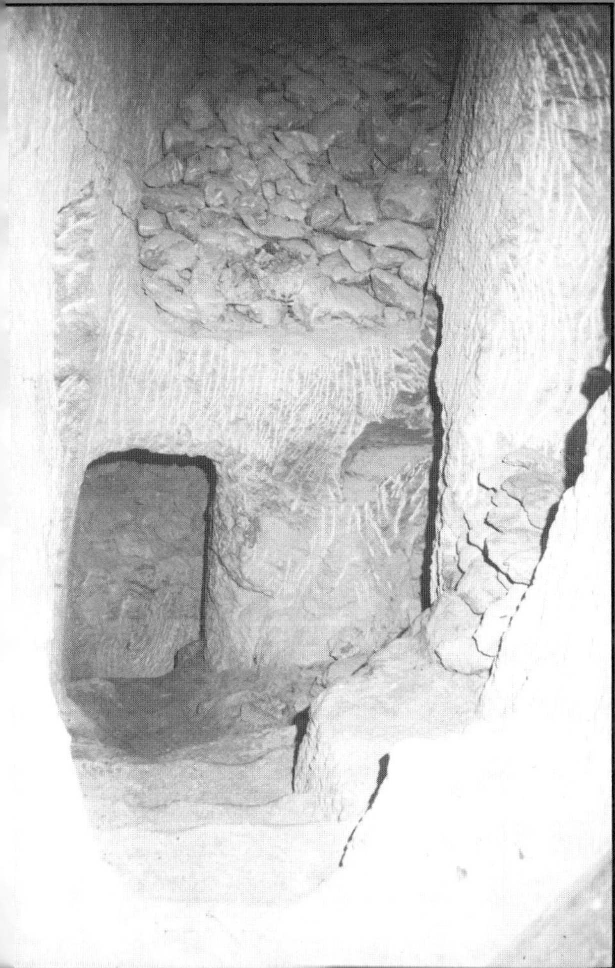

302 Torrens No. 1 talc mine in 1975, viewed down a stope. Steps in the foreground have been cut for access to lower levels; several areas have been backfilled with waste rock. White scratches on the walls are pick marks.

Tumby Bay

Mining began on a series of small irregular lodes in 1909. Output of high grade white talc, selectively mined mainly from small underground workings serviced by more than fifty shafts, reached a peak during 1941-48. The underground operations ceased in 1956, but some talc has been obtained by intermittent open cut activity.

303 Bagged talc being hoisted from a shaft at one of the Tumby Bay mines in 1927.

304 Tumby Bay talc mines in 1944. The line of lodes is outlined by spoil from the numerous shafts. Twelve thousand tonnes of talc were mined here up till 1956.

Mount Fitton

The extensive Mount Fitton deposits were discovered by J. G. Ford in 1945, and mining has since been continuous.

Approximately forty lodes have been located over an area of about sixty square kilometres, with some containing several hundred thousand tonnes of talc. About half of these have been worked.

Most of the mining has been by open cut, but one of the deposits was worked from underground between 1956 and 1967.

Total production to the end of 1980 was two hundred thousand tonnes; current production is about twelve thousand tonnes per year.

305 An exploratory pit near Mount Fitton in 1946.

306 Talc being barred down from the face at Mount Fitton Mine in 1946. Workmen in the foreground are hand trimming stained portions.

307 Inclined railway from an open cut to the trimming area at right.

308 Open cut and underground mining methods have been used at Mount Fitton; a shaft collar is in the floor of the opencut at lower right. Loaded ore skips were hauled up greased wooden rails from the shaft to unloading facilities in the centre.

309 *An open cut at Mount Fitton in 1968.*

310 A picking bench, where badly stained talc was knapped by hammer and minor iron staining removed with wire brushes. Pure white lumps of talc were then bagged and transported to Adelaide for milling. Picking benches are no longer used at Mount Fitton.

311 Another method of upgrading run-of-mine talc. Stained portions are being trimmed with a jack pick, and selected high grade material is being loaded into the ore bin.

312 Lower grade talc is crushed, screened and sorted for general industrial use.

GYPSUM

South Australia has long been an important producer of gypsum, and now provides over two thirds of the Australian output which totals almost one million tonnes annually.

The most important deposits are at Lake Macdonnell, Lake Fowler, Stenhouse Bay, Salt Lake and Pelican Lagoon (New Lake) on Kangaroo Island, Cooke Plains and Blanchetown. Large unworked deposits occur at Streaky Bay and Parcoola.

Calcined gypsum, as plaster of Paris, is used in the manufacture of plaster sheets, wallboards, ceilings, mouldings, and as a fireproof lining to protect structural steelwork. Raw gypsum is added to portland cement to retard setting time, and as finely ground powder is used in fertilizers as a soil conditioner, and as a filler in paper, textiles, paint and rubber.

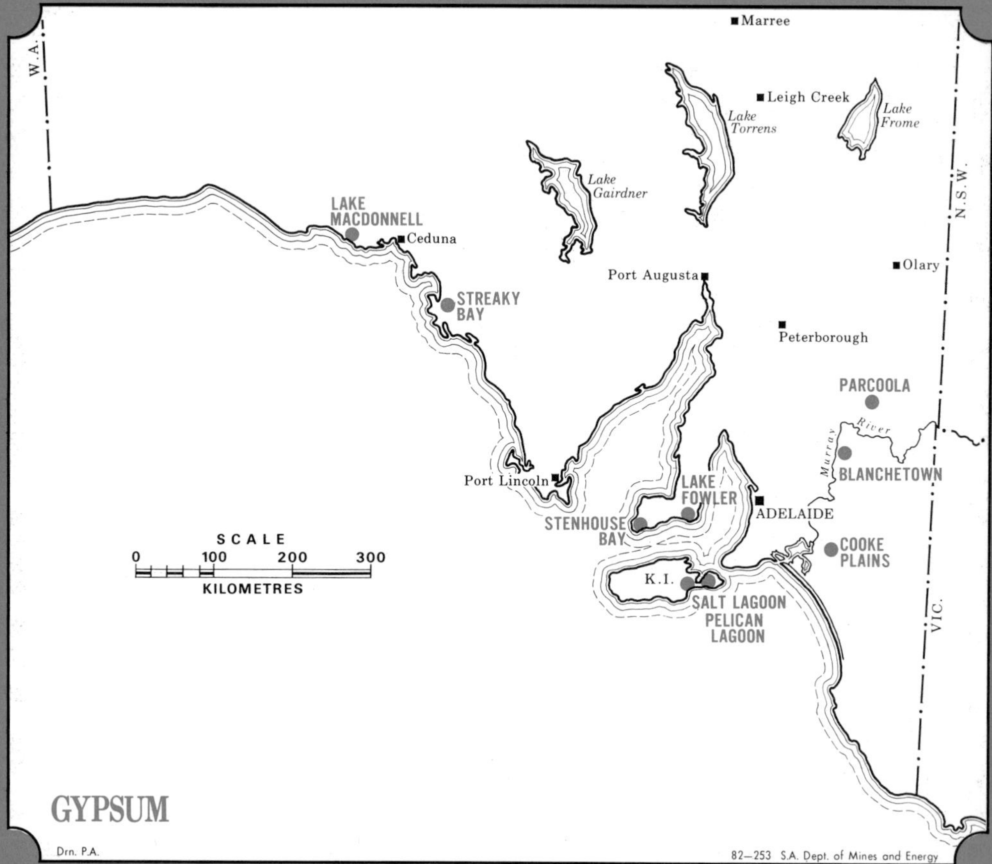

Lake Fowler

The first leases were pegged in 1874, but mining did not commence until 1905 when a large dune of gypsum sand on the lake shore was excavated. This yielded over three quarters of a million tonnes to the end of 1980.

At present, Adelaide Brighton Cement Ltd produces a few thousand tonnes annually for cement manufacture.

313 Gypsum mining at Lake Fowler in 1915. A large dune of gypsum sand, which extends for over one kilometre along the lake shore, was mined by pick and shovel and bagged on site for transport to Edithburgh. Production has been continuous since 1905.

314 Lake Fowler workings in 1950. Gypsum blasted from the dune face is being hand sorted before loading and transport. All produce at this time was shipped from Edithburgh to New Zealand for cement manufacture.

Stenhouse Bay

In the 1890s, Australian Gypsum and Whiting Co. Ltd erected a tramline, jetty and loading facilities at Marion Bay to handle gypsum mined from Marion Lake. A plaster factory was built in Melbourne, but the company went into liquidation before operations commenced. It was not until 1905 that a small amount of gypsum was bagged and sold.

Permasite Manufacturing Co. Pty Ltd, known later as Peninsula Plaster Co. Pty Ltd, pegged leases over Inneston Lake in 1913, constructed a washing, crushing and plaster works on the lake shore, and erected a jetty at Stenhouse Bay. The lake provided the bulk of South Australian production until 1930 when Waratah Gypsum Pty Ltd was formed by a consortium of southern Yorke Peninsula gypsum miners. From that time, Marion Lake became the only producer in the area.

Over six million tonnes of gypsum have been taken from Inneston and Marion Lakes, the latter being almost worked out.

316 At Inneston Lake on Yorke Peninsula, rock gypsum occurs as a one metre thick crust. Gypsum broken from the working face at left has been stockpiled for transport to the jetty at Stenhouse Bay.

315 Inneston Lake and plaster works of Peninsula Plaster Co. Pty Ltd in the early 1920s. Gypsum mined from the lake was crushed, washed and calcined to produce plaster. A large pump lowered the water table below the lake, enabling operations to be continuous throughout the year. Workmens' houses are to the right of the factory.

317 Loading facilities at Stenhouse Bay in 1935. Gypsum was hauled by rail from Marion Lake to the plant at Stenhouse Bay where it was crushed, washed and screened before storage in the cliff-side bin above the jetty. A conveyor belt from the bin to the end of the jetty loaded gypsum directly into steamships bound for the eastern States. Workmen are shovelling gypsum towards the chute during loading.

Lake Macdonnell

Two companies currently supply half of the Australian market from the lake, the largest worked deposit of gypsum in Australia.

Gypsum, first produced here in 1919, was bagged and carted by horse or bullock wagon to Port Le Hunte for shipment to Sydney. On completion of the Thevenard to Penong railway in 1923, Port Le Hunte was abandoned in favour of Thevenard where a plaster factory was erected.

Production ceased in 1930 due to increased output of lower cost gypsum from Stenhouse Bay, and recommenced in 1947 under Waratah Gypsum Pty Ltd, the current operators. Colonial Sugar Refining Co. (CSR) has mined gypsum sand from dunes near the lake since 1948. Current annual output totals almost a half million tonnes.

Total production amounts to about seven million tonnes.

318 Gypsum mining at Lake Macdonnell during 1920-23. Fifty men were employed to break, bag and load gypsum into horse-drawn wagons. From 1920 to 1922, wagons carried bagged rock to Port Le Hunte. The railway from Kowulka Siding to new plaster works at Thevenard was completed in 1923, but horses were required to cart gypsum to the siding before erection of the now abandoned aerial ropeway.

319 Gypsum loading facilities at Lake Macdonnell in 1928. The aerial ropeway to Kowulka Siding replaced horse-drawn wagons. It was eight kilometres long, had an operating speed of eight kilometres per hour, and a capacity of sixty-five tonnes per hour. A railway spur now connects Lake Macdonnell to Thevenard.

322 Stockpiles and railway loading facilities at Lake Macdonnell.

320 Blasting hard rock gypsum at Lake Macdonnell. A blasthole drill and air compressor are in the foreground.

321 Dragline stockpiling gypsum which has been dredged from shallow water at the mining face.

323 Bulk gypsum loading facilities at Thevenard.

Salt Lake

Operations commenced at Salt Lake (Kangaroo Island) in 1958 when raw gypsum was trucked to American River for shipment by shallow-draft ketches to the plaster factory at Port Adelaide.

CSR took over operations in 1959, erected a crushing and washing plant at the lake, and installed bulk ship-loading facilities at Ballast Head. Three million tonnes have been mined from Salt Lake, and nearly three hundred thousand tonnes are now mined annually.

Deposits located at Pelican Lagoon (New Lake) are being prepared for mining as the Salt Lake reserves are almost depleted.

324 Gypsum mining at Salt Lake on Kangaroo Island in 1960. Rock gypsum was drilled, blasted, and loaded by backhoe into trucks for delivery to the lakeside crushing and washing plant.

325 Gypsum mining at Salt Lake in 1980.

326 The nearly worked out Salt Lake gypsum deposit. Forty thousand tonnes of reserves, approximately two months supply, remain in the foreground.

327 Salt Lake washing plant. Stockpiles of processed gypsum are to the left and at upper right; these are allowed to stand for up to a year, during which time the more soluble contaminants are leached from the gypsum.

328 The spiral screw gypsum washing troughs at Salt Lake.

329 Ballast Head stockpiles and loading facilities. Three hundred thousand tonnes of gypsum, representing a full year's mining, are contained in the stockpiles.

PHOSPHATE

Rock phosphate was discovered near Clinton in 1899, and later at Kapunda, Moculta, Myponga, and in the Flinders Ranges. Desultory mining operations have been undertaken at these deposits since 1903, and four hundred thousand tonnes have been produced.

Phosphate from Kapunda and Myponga was used during the Second World War by The Broken Hill Proprietary Co. Ltd for manufacture of high phosphorus pig iron. Only the Moculta and Kapunda deposits are now worked. The current production is finely ground and applied directly to the soil as a fertilizer in areas of high rainfall.

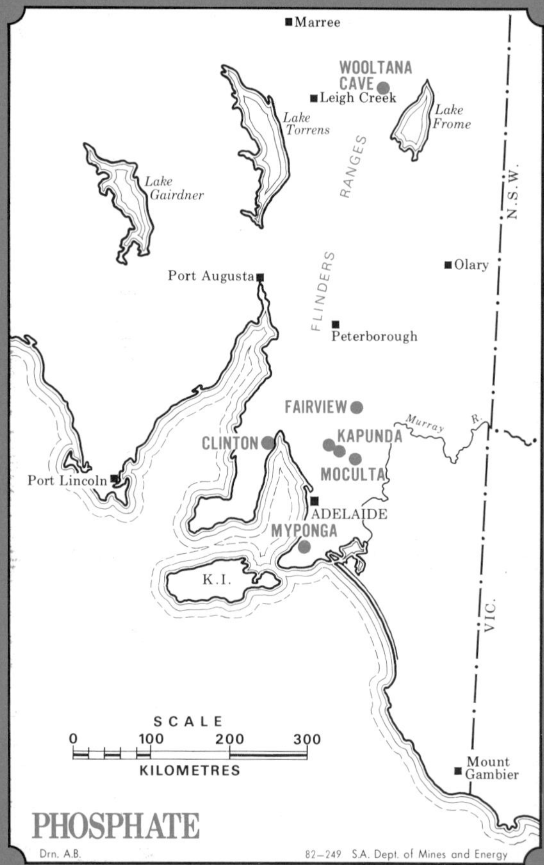

330 Fairview phosphate quarry, twenty-six kilometres southeast of Burra, at about 1907. This was the second South Australian rock phosphate discovery after Clinton. Workers employed by the Adelaide Artificial Manures Co. are loading rock via a chute into a horse dray on the middle of three benches; an exploratory adit is at left.

331 A rock phosphate quarry near Kapunda, circa 1910. The rock was mined with pick and shovel, stockpiled in heaps at right, and hand sorted to remove waste.

332 Green's phosphate quarry near Kapunda at about 1918. This deposit was discovered in 1904, one year after the Fairview phosphate. Three quarries currently working in this area produce eight thousand tonnes per year.

333 Workers employed in the Kapunda rock phosphate quarry shown in the previous photograph. Vests and wide-brimmed hats were considered to be part of the miners' attire, and two miners in the front row are wearing knee straps. The men at extreme left and right were probably company clerks.

334 Wooltana Cave guano deposit in 1924. The cave is located in a dolomite ridge ten kilometres northwest of Wooltana Homestead in the northern Flinders Ranges. The guano contained ammonium, potassium and calcium salts derived from bat droppings and remains. Bagged guano was hauled up a slide from the main chamber, over sixty metres from the entrance, to the wooden headframe. It was treated with sulphuric acid in a similar manner to rock phosphate for production of superphosphate.

BARITE

South Australia is the most important source of barite in Australia. Before 1945, most was produced from mines near Noarlunga, but since then the main source has been Oraparinna Mine in the Flinders Ranges.

Barite is a dense mineral used in drilling muds to contain high fluid pressures in deep wells, as white pigment in paint and quality paper, as a flux in glass manufacture, and as a filler and extender in rubber, paint, textiles and linoleum. It is also used in medicine, and as aggregate in walls around x-ray equipment.

The first mines in the State were opened in 1910 at Williamstown, and on the site of the present Aldgate Recreation Reserve in 1913. Other small deposits were worked near Glen Osmond, Uraidla, and north of Birdwood.

The most important early discovery was south of Noarlunga in 1918, where three underground mines were opened; the largest, Crittendon, was worked to a depth of seventy metres. By 1935, seven mines were in operation with much of the work being done by tributers. The mines closed in 1961 after sixty-five thousand tonnes had been produced.

Prospectors moved progressively northwards from Adelaide after the Noarlunga find, firstly exploring deposits in the Mount Lofty and Flinders Ranges, and then the Olary region and Eyre Peninsula. Small lodes were worked near Truro, Whyalla, Corunna, and Pernatty Lagoon; large low grade deposits were found at Walparuta, Dome Rock, and Mount Mulga.

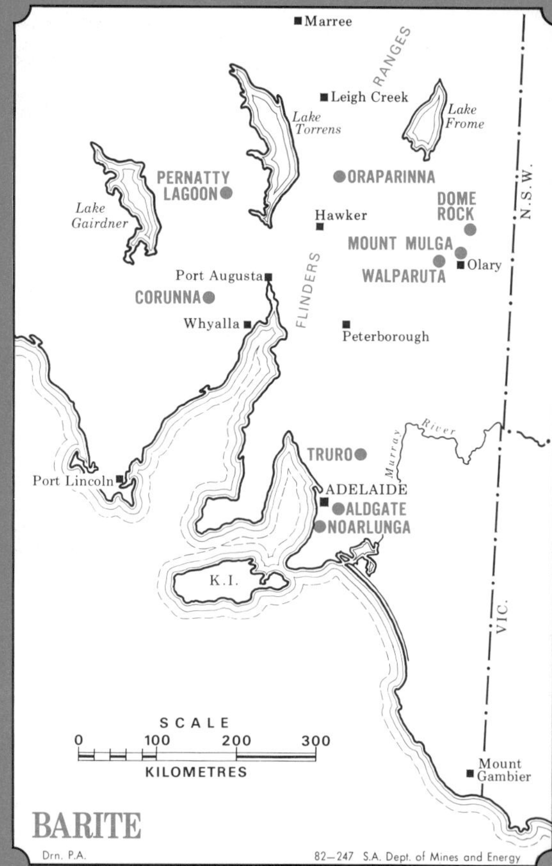

335 Aldgate barite mine in 1916. Ore was carried to the portal of the two adits by rail trucks for hand trimming to remove waste or stained portions; a trimming area is in the right foreground. Selected barite was carried by horse-drawn sled to the nearby crushing and grinding plant. About five thousand tonnes of graded barite were produced from 1913 to 1921.

336 Dutton barite mine, near Truro, before closure in 1928. The open cut is on a steeply dipping vein. Over one thousand tonnes of high purity translucent barite were trucked to Truro and railed to Adelaide.

Oraparinna

The most important discovery of barite in Australia was made by J. Tobin at Oraparinna in 1940. Early production was from an open cut, where the ore was hand sorted to reject stained and undersized pieces, and lowered by flying fox (later an inclined tramway) to a bagging and loading bay.

Underground mining commenced in 1948, and by the mid-1950s large volumes of stained yet high grade barite had been accumulated at the mine. With assistance from the South Australian Government, a treatment plant was established at Quorn in 1957 to produce a clean milled product.

Expanded petroleum drilling activity since the mid-1960s has occasioned a greatly increased demand for barite from Oraparinna and nearby smaller lodes.

The 1940-80 production from Oraparinna and adjacent lodes totalled almost two hundred and fifty thousand tonnes.

337 Oraparinna open cut in 1940, a few months after the mine was opened. Hand picked ore is being bagged for transport to Adelaide.

338 Bagged ore from the open cut was lowered halfway down the hill by flying fox to a loading bay.

339 By 1958, ore was being dropped down chutes to a wooden storage bin where it was sorted, bagged, and lowered down the inclined tramway to stockpile and loading facilities in the foreground.

340 Oraparinna Mine in 1980. The mine has been developed on a series of steeply dipping barite veins in The Bunkers Range of central Flinders Ranges. The original open cut is a dark shadow at left centre above the adits leading to the underground workings of the main mine. Scree slopes of mullock have formed below open cuts on other lodes at right.

341 A barite lode after mining, in 1980. The miner is standing on a pillar of white barite left in place to maintain wall stability.

342 Linke's Lode, fifteen kilometres southwest of Oraparinna Mine, circa 1940. Three light-coloured barite veins crop out on the hill in the middle distance. The car is on the main Wilpena Pound-Blinman road.

343 A similar view in 1981. Iron-stained barite for oil-well drilling mud is stockpiled in the foreground.

344 Barite mill and treatment plant at Quorn in 1965. Ore is carried by road to the mill from Oraparinna and other deposits in the Flinders Ranges. The milled product was railed to Adelaide until the line through Pichi Richi Pass closed; road transport is now used. The heavy nature of barite is demonstrated by the half-filled yet fully-laden rail wagons.

ALUNITE

Alunite deposits were investigated during the First World War when potash supplies from France and Germany were curtailed. Mining was undertaken on cliff tops north of Carrickalinga Head from 1913, and three hundred and fifty tonnes were shipped from the Normanville jetty until closure of operations in 1919.

Further discoveries were made in 1914 at Napperby, from which forty tonnes were mined until 1917, and near Stansbury in 1918, from which two hundred tonnes were extracted until 1924. Other occurrences are at Coober Pedy, Rapid Bay, Port Vincent, Angepena, Nepabunna, and southwest of Oodnadatta. The large Pidinga deposit was discovered in 1948.

Alunite has been used as a source of potash for fertilizers, and as an ore of aluminium.

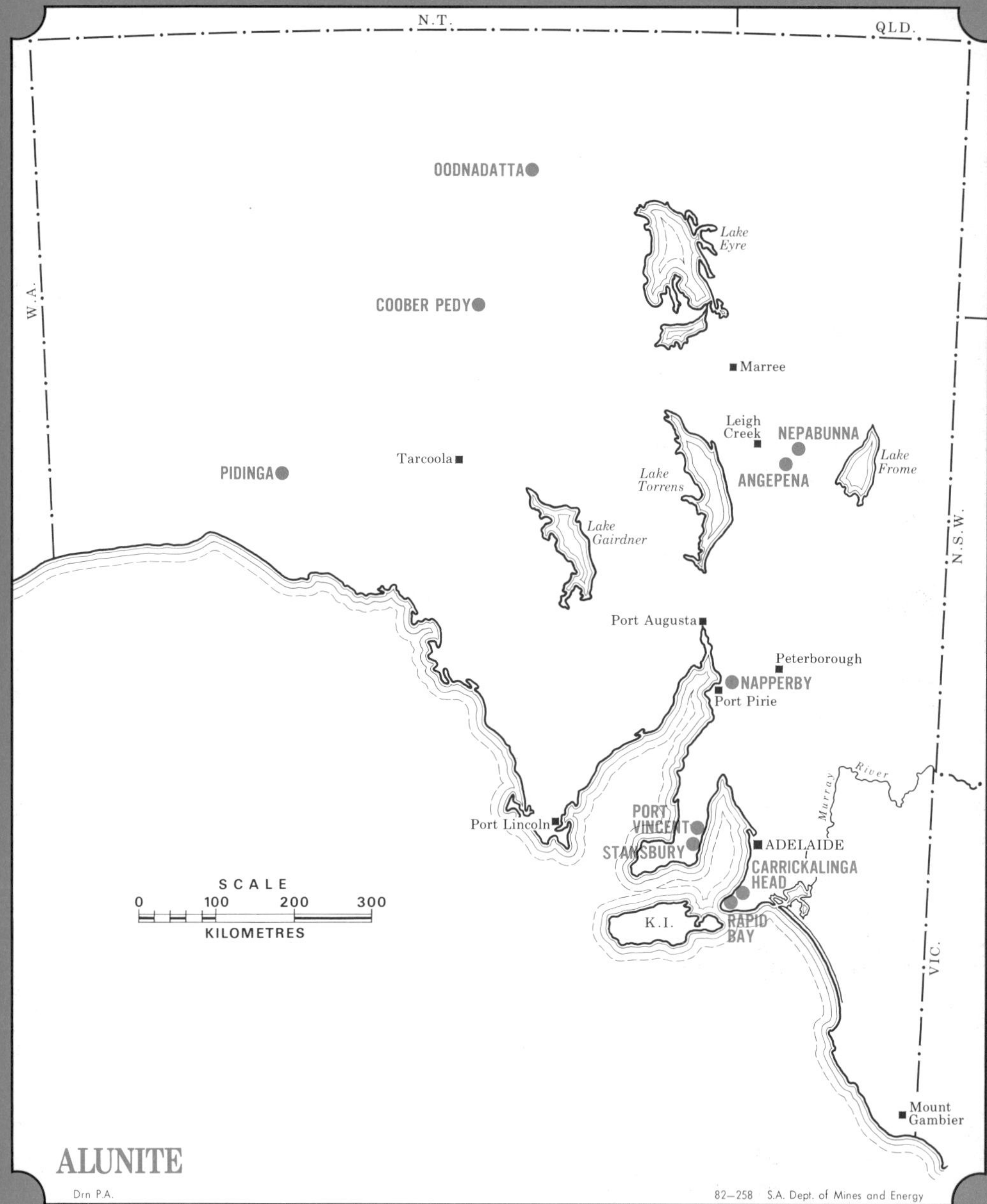

345 Adit and shaft at Napperby alunite mine in May 1917.

346 Carrickalinga open cut at about 1918. Following discovery in 1913, a small trial sample of alunite was sent to England by an Adelaide syndicate. After a favourable quality report, Carrickalinga Alunite Co. was formed in 1914 and began mining from open cuts, adits and shafts.

FELDSPAR

Mining began in 1909 at Chinatown, near Hog Bay (now Penneshaw), on Kangaroo Island. The Kangaroo Island China and Clay Co. recovered feldspar, chinastone (an intergrowth of feldspar and quartz), and fireclay from a pegmatite that had been previously worked for gem quality tourmaline; firebricks and pottery were manufactured at Hog Bay.

The main source of feldspar in South Australia was a pegmatite two kilometres east of Gumeracha, which produced potash feldspar and a mixture of potash and soda feldspar for the manufacture of glass and pottery. The quarry was worked continuously from 1933 until closure in 1969, firstly by the South Australian Feldspar Co. N.L., and later by the S.A. Silicates Co. Pty Ltd.

The other major source of feldspar has been a group of pegmatites north of Olary, four hundred kilometres northeast of Adelaide, which were worked between 1932 and 1979. These pegmatites have also been the State's only commercial source of beryl, mined as an ore of beryllium metal from 1941 to 1963.

Minor quantities of feldspar have been mined at Williamstown, Lobethal, Barossa, Myponga and Cambrai.

347 ***Cold Tea feldspar mine at Olary in 1936. Broken feldspar was winched to the surface by the hand windlass and bagged for transport. The feldspar was worth £2/18/- per tonne on rail at Olary.***

348 ***White Rock feldspar and beryl mine at Olary in 1954. Harry and Nell Lively are hand picking beryl crystals from gangue. The largest beryl crystal found is in the foreground.***

349 ***Raven Hill feldspar mine at Olary in 1942. A power winch is being used to haul a side-tipping wagon of feldspar out of the open cut. The wagon will be turned at right angles on the flat plate and railed downslope to the loading area.***

OPAL

Approximately eighty per cent of the world's production of opal comes from the South Australian opal fields.

Even before opal was discovered at Coober Pedy in 1915, Australia was the world's leading supplier, most of which came from deposits at White Cliffs (Qld), and Lightning Ridge (NSW). Mining began at Coober Pedy towards the end of the First World War and accelerated with the migration of ex-soldiers to the opal fields. They introduced the dugout style of living in a successful attempt to combat the prevailing extremes of temperature.

A severe drought and depressed markets affected the industry during 1922-24, as did the depression of the 1930s when few people could afford to buy luxuries such as opal. However, in 1946, major new finds stimulated activity; since then a spectacular increase in opal production has developed into a multi-million dollar industry.

Unlike other mining industries in South Australia, opal miners are not required to report production, and official figures relating to value are estimates only. Since 1978, the value of opal produced is estimated by the Department of Mines and Energy to be nearly $40 million annually.

Opal occurs in deeply weathered rocks to depths of forty metres. Methods of mining in the early days were simple and inexpensive, with prospectors' tools comprising a pick, shovel, pocket knife, and pliers. Workings consisted of shallow trenches only one or two metres deep. As the shallow ground was worked out, the miners went deeper and windlasses were used to haul mullock from the diggings. Mining methods have become increasingly mechanised over the years, and few miners still mine solely by hand.

Modern equipment, such as bulldozers, tunnelling machines, and blowers, is now extensively used but the traditional handpick, screwdriver or pocket knife is still used to prise out the opal.

Even noodling, the process of sorting through mullock heaps to extract overlooked opal, has become mechanised.

Opal production in South Australia is almost entirely confined to the three main opal fields of Coober Pedy, Andamooka, and Mintabie. A small field at Stuart Creek, north of Andamooka, has also been worked intermittently since its discovery in 1947, and there are numerous other small opal diggings across the north of the State.

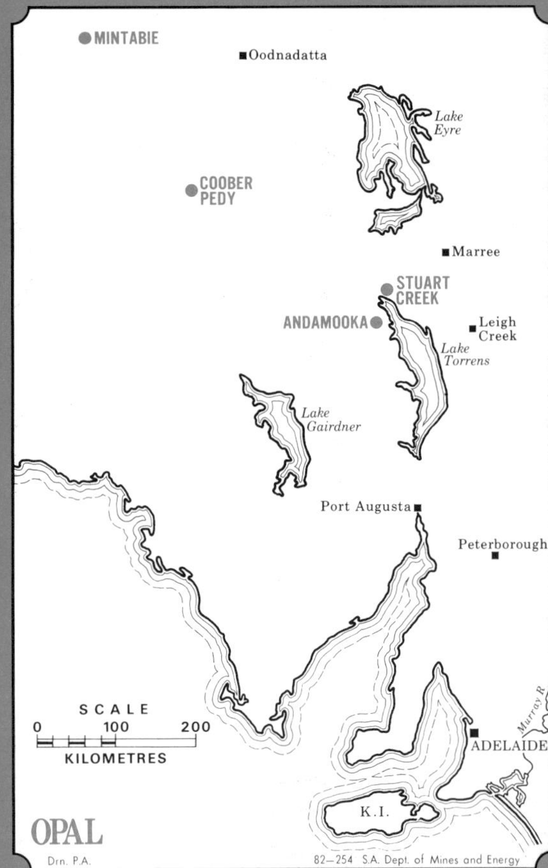

Coober Pedy

At the present time more opal is produced from here than all other Australian fields combined. The discoverer was fourteen-year-old Willie Hutchison who was accompanying his father in a gold prospecting party. Willie had been left to look after the camp while the men searched for water. They returned after a futile effort to find the camp deserted. Willie strolled in just after dark and announced that he had found several pieces of opal. The party's original object was forgotten, and many brilliantly-coloured pieces were found before lack of water forced them to leave the area.

The district was first known as the Stuart Range Opal Field but the name was changed to Coober Pedy, aboriginal for 'white man's burrow', in 1920. Early workings were shallow and tended to follow the Stuart Range escarpment. Deep mining did not begin until 1946 with the discovery of opal at Geraghty Hill, west of Coober Pedy township.

Much of the opal produced is milky or white, and of lower value than that from either Mintabie or Andamooka. Some of the opal is of very high quality, however, and nearly all finds a ready market, particularly amongst Chinese buyers.

350 Trying out new ground at Coober Pedy in the 1920s. The miner, having found opal floaters on the surface, is sinking a shallow shaft. This photo was probably taken in the Big Flat area, later to become the principal field of the time.

351 Coober Pedy Post Office and store in 1925. The post office was established in 1920 at Post Office Hill and is now part of the Miners Store.

352 Aerial view of portion of the Coober Pedy Opal Field in 1979. Bulldozer cuts in the foreground are surrounded by exploratory auger holes. Opal fields expand by peripheral growth as new finds are located adjacent to the mined areas.

353 Auger drill at Coober Pedy. These truck-mounted rigs are capable of boring a small diameter hole to a depth of forty metres. An auger drill is generally used to prospect a claim prior to open cut or underground mining.

354 Electrically powered tunnelling machine at Coober Pedy in 1977. The cutting head, mounted on a slowly rotating arm, excavates a circular tunnel. Mullock is transferred to the surface by a blower; the miner's foot is resting on the blower pipe.

355 Another type of tunnelling machine at Coober Pedy in 1981. The rotating cutting head, equipped with hardened steel teeth, are mounted on a boom that moves slowly up and down to cut a square opening. In favourable rock conditions, this machine can advance twenty metres per day.

356 A blower at Coober Pedy in 1978. These truck-mounted machines operate like a vacuum cleaner. A fan, or blower, driven by a stationary diesel engine, draws mullock out of the shaft through a series of metal pipes. Blowers are sometimes combined with a conveyor system to dump waste material beyond the shaft.

357 Underground miners carefully gouging opal from a level on the Olympic Field, Coober Pedy, in 1981. The miner is holding his hand under the vein to prevent the loss of small chips.

358 Noodling machine for sorting opal chips from mullock on the Black Flag Field at Coober Pedy in 1979. Mullock is fed by a front-end loader into the hopper at left, from where it passes through the revolving screen to remove undersize material and into the box-like darkroom. Here ultra-violet light causes most of the opal to fluoresce, enabling it to be hand picked. Waste material is discharged by the conveyor at right.

359 Aerial view of Coober Pedy in 1981 looking easterly towards the town. The Stuart Highway connecting Adelaide and Alice Springs is at centre left. Bulldozer cuts and auger holes are in the foreground.

Andamooka

Andamooka Opal Field, famous for its crystal opal, was discovered in 1930 by Sam Brooks and Ray Shepherd.

Andamooka was a major producer of precious opal until 1972, with output equalling in value and occasionally exceeding that from Coober Pedy. Despite the discovery of several small fields, the failure to locate a major new find since 1971 has resulted in a steady decline in production and population. Several finds in the White Dam area in late 1979 stimulated activity again, and development of the nearby Roxby Downs copper-uranium-gold deposit may increase the number of part-time miners.

Opal at Andamooka is scarcer than that at the other fields, but is generally of good quality and is eagerly sought by buyers.

360 Westerly view of Andamooka township from One Tree Hill in 1948.

361 Andamooka opal workings in 1948. Mullock dumps have been built up around the shafts, some of which are fitted with hand windlasses for hauling waste rock to the surface. Hand windlasses have now all but disappeared from the fields.

362 *Aerial view of Andamooka township in 1969. Shafts and bulldozer cuts of German Gully Field are in the foreground.*

363 Calweld drill sinking an exploratory shaft at Andamooka in 1976. These rigs are capable of drilling one metre diameter holes to depths of twenty-five metres. The bucket is lowered down the hole and rotated like a post-hole digger. When full, it is brought to the surface and mullock dumped away from the hole to be searched for traces of opal.

364 The self dumper, or automatic bucket tipper, allows the opal mining operation to be controlled by one miner from underground. In the background is a petrol-driven Yorke hoist which requires an operator at the surface to hoist and unload the bucket.

Mintabie

The original opal discovery at Mintabie is not recorded but the find has been attributed to Larry O'Toole during the early 1920s.

Because of remoteness, lack of water, and hard ground, the field has been worked intermittently. Large bulldozers were used in late 1976, and several major finds resulted in a considerable increase in activity. The prohibitive cost of operating large bulldozers in the hard sandstone has forced many of the large machines from the field, but the number of miners has remained reasonably constant as they have resorted to underground mining.

365 Aerial view northeasterly of Mintabie Opal Field in 1979. The large mullock heaps are the result of bulldozing operations on the edge of the escarpment.

366 Open cut mining at Mintabie in 1978. Bulldozers are ripping and pushing sandstone out of the cut while spotters are watching for signs of opal.

URANIUM AND RADIUM

At the end of the Second World War, South Australia was considered to possess the most prospective uranium deposits in the British Commonwealth, at Radium Hill and Mount Painter. After extensive exploration, the Radium Hill Mine was commissioned in 1954. In seven years of operation, the mine produced nearly one million tonnes of ore, from which uranium was extracted as yellowcake at a treatment plant in Port Pirie.

The Mount Painter deposits, some of which had been mined for radium in the 1920s, proved to be of too low grade for economic recovery of uranium. Further deposits have since been located.

A small quantity of uranium ore was produced from Wild Dog Mine near Myponga during 1955-57 and treated at Port Pirie.

Uranium deposits with potential for development occur at Olympic Dam, Beverley, Honeymoon, Goulds Dam and Yarramba. Small deposits are known at Crockers Well, Mount Victoria, Port Lincoln, and Mount Shanahan.

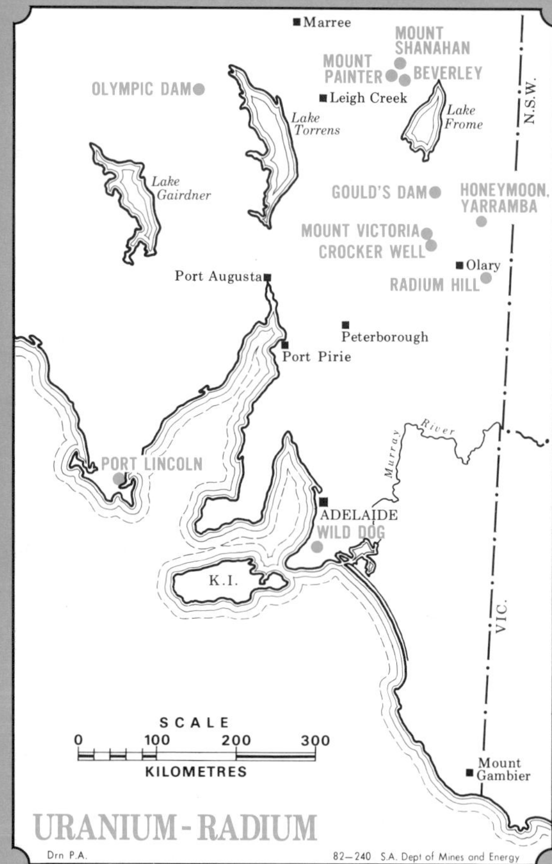

Radium Hill

Uranium mineralisation was discovered at Radium Hill in 1906 by prospector A. J. Smith who at first thought the heavy dark rock to be an ore of tin.

The Radium Hill Co. took over the claim, and six hundred tonnes of ore were mined by 1911. Other companies worked without success, and by the end of the First World War five shafts had been sunk and abandoned. The ore concentrate was treated in New South Wales and Victoria to yield a few hundred milligrams of radium and several hundred kilograms of uranium by-product.

The Radium and Rare Earths Co. erected a plant at Dry Creek, near Adelaide, in 1923 to produce radium and radium bromide from the Radium Hill ore, but this proved to be uneconomic by 1932.

In 1944, the Commonwealth Government received a request from Britain for assistance in evaluating Australian uranium provinces. The Second World War ended before the program was completed, and the South Australian Government took over the project in this State.

By 1952, Radium Hill had been proved large enough to mine, and many problems associated with ore beneficiation had been solved.

A town to accommodate eleven hundred people was built, water piped in from New South Wales, a railway spur constructed to connect Radium Hill with the Broken Hill-Port Pirie line, and a power transmission line constructed from Morgan. The mine was officially opened on 10 September 1954 by the Governor-General, Field Marshall Sir William Slim.

A new main shaft was sunk to more than four hundred metres, and output eventually reached six hundred tonnes per day. After crushing and concentrating on site, the concentrate was sent to Port Pirie for treatment.

Completion of overseas contracts and exhaustion of reserves led to the mine closure in December 1961.

367 The mine and concentrating plant at Radium Hill in 1926. The plant was erected by the Radium and Rare Earths Treatment Co. over the old workings. An attempt was made to recover by-products, such as titanium white used in paint pigments, but the complex ore prevented economic extraction and operations ceased in 1931.

368 Southeasterly view of Radium Hill township in 1957. The mine lies about three kilometres behind the large water tank on the hill.

369 Official starting of the Radium Hill Mine ball-mill on 10 September 1954. After crushing, the ore passed into the ball-mill for fine grinding. Flotation processes were used to separate davidite, the main uranium mineral, from the gangue; davidite was extracted from the resultant slurry by disc filters which can be seen at top left. The ball-mill was operated above design capacity to cope with high ore production rates from the mine.

370 Miners drilling blast holes with a jackhammer in 1955. A carbide lamp hangs from the stope back.

371 Mucking out blasted ore with an ore truck on inclined rails; these were used to minimise removal of waste rock during the mining of narrow lodes.

372 An air-driven rocker shovel for mucking out ore in the larger stopes. The miner is wearing a battery-powered cap lamp, the most common present day form of personal underground lighting.

373 Ore being loaded into railway wagons on a level in 1954. Broken rock was dropped down ore chutes from stopes above the level to this centralised loading station, one of many in the mine. A small diesel locomotive in the background hauled ore to the main shaft for hoisting by skips to the surface.

374 Headframe over the four hundred metre deep main shaft in 1954.

375 Aerial view of Radium Hill Mine in 1957. The main shaft is in the right foreground alongside the mill and concentrating plant. The large square structure in the left background is a tailings dump built up of waste material from the concentrating plant.

376 A 1958 photograph of the uranium treatment plant, built by the State Government, at Port Pirie. The plant processed all output from the Radium Hill and Myponga mines. Ore concentrates were leached in boiling sulphuric acid to dissolve the uranium, which was precipitated, after further processing, as yellowcake. Leaching vats were housed in the tall building to the right of the four thickening tanks. The tanks were used to settle out solids from the waste material, and water was returned to the plant for re-use; the thickened solids were pumped into the tailings dams at left. During operations from 1955 to 1962, the plant produced over eight hundred and sixty tonnes of yellowcake which was sold to the British and USA governments.

Mount Painter

Uranium bearing minerals were discovered in the Mount Painter area by G. A. Greenwood, son of a local pastoralist and prospector, W. B. Greenwood, in 1910. This discovery, on what was later named Radium Ridge, was exploited for radium by the Radium Extraction Company of South Australia Ltd.

The company opened up several other deposits during the next two years, the largest being the No. 6 workings near Mount Painter. Ore was shipped to Europe until the outbreak of war in 1914; the company went into liquidation in 1917.

The workings were reopened in 1923 by what was to become the Australian Radium Corporation N.L. Lack of water prevented the erection of a leaching plant on site, and a small crushing and screening plant gave unsatisfactory results. Camels were used in the rugged terrain to carry crude concentrate and hand-picked ore to less hilly country, where it was transferred to motor lorry and carted to the railway at Copley. From there it was railed to the Dry Creek treatment plant. The Australian Radium Corporation ceased operations in 1932.

Exploration at Mount Painter resumed in 1944 in conjunction with re-examination of Radium Hill and other uranium deposits. Roads were constructed into the workings and camps erected, but the deposits proved to be low grade and uneconomic. Drilling and underground development was continued at East Painter by the South Australian Government until the project was abandoned in 1950.

Between 1968 and 1971, a consortium of mining companies discovered further uranium deposits including the richest and most easily accessible at the Hodgkinson Prospect.

***377** Camels being loaded with hand selected ore at the No. 6 workings, near Mount Painter, at about 1912. The ore was carried west along the camel pad at the base of Radium Ridge to Blue Mine Creek for motorised transport to the railway at Copley. The dark outcrop at top left is mainly hematite, and contains the uranium minerals.*

378 Transferring supplies from motor lorry to camel near Blue Mine Creek in 1926. The lorry carted ore and concentrates to Copley for railing to the Dry Creek radium extraction plant.

379 No. 6 workings and concentrating plant at about 1930. The ore crops out at upper left, with a small open cut and mullock heap to the right. The main ore chute leads to a wooden storage bin just above the concentrating plant roof.

380 Concentrating plant at the No. 6 workings in 1926. Bagged ore was dropped from the mine to the plant down a chute at top left. One of the workers is emptying ore into a small belt-driven jaw crusher; the crushed rock was fed into the ball-mill at right for grinding prior to sieving. The product was sent to the treatment plant at Dry Creek.

381 *Camels at Mount Painter Camp in 1944. Radium Ridge, site of the original uranium discovery, is in the background. Camels were preferred to horses owing to the lack of permanent water in the area. A well which produced good quality water was sunk a few metres from the eucalypt tree at extreme right.*

382 *A 1947 view of East Painter Camp which accommodated up to sixty workers and was equipped with a canteen, bunk-houses, and garages. It was erected in 1944 and dismantled in 1950.*

383 Department of Mines diamond drilling at East Painter in 1947; East Painter Camp is behind the trees.

384 Government geologists engaged in mapping and prospecting at Mount Painter in 1947. The prominent hill in the background is The Armchair.

385 Mullock being dumped at the mouth of an East Painter exploratory adit; Mount Painter is in the background.

386 Official visit to the East Painter Camp in 1947. Mark Oliphant, the atomic physicist (later Governor of South Australia), is seated fourth from left next to the South Australian Premier, Thomas Playford. The Governor, Sir Willoughby Norrie, is seated to the Premier's left, and seated at extreme right is Ben Dickinson, Director of the Department of Mines.

387 The Mount Gee uranium prospect near Mount Painter in 1970. Approximately one hundred exploration holes were drilled into this prospect between 1969 and 1971.

388 A percussion drill rig at the Mount Gee uranium prospect; Mount Gee is in the background.

PYRITE

A world shortage of sulphur in 1951, and the prospect of limited future supplies, led to the establishment of a company to develop deposits of pyrite (iron sulphide) at Brukunga near Nairne. Nairne Pyrites Ltd was formed by three South Australian fertilizer companies—Cresco Fertilizers Ltd, The Adelaide Chemical and Fertilizer Co. Ltd, and Wallaroo-Mount Lyell Fertilizers Ltd—in association with Broken Hill Proprietary Co. Ltd. The undertaking was sponsored by the South Australian Government which conducted early investigational work through the Department of Mines. The Commonwealth Government guaranteed to pay a subsidy if the market price of sulphur fell below an agreed price.

The quarry, crushing and treatment plant began operation in early 1955, and the first consignment of concentrate was despatched a few months later to the Birkenhead plant of Sulphuric Acid Pty Ltd for processing into sulphuric acid for superphosphate manufacture. The township of Brukunga was established to house the quarry employees; road, rail, power and water services were supplied to support the mining operations.

About three hundred thousand tonnes of pyritic ore were mined annually by open cut. The ore was crushed in jaw and cone crushers, then ground in rod and ball mills. Concentration by flotation resulted in a product containing 40 per cent sulphur. After filtration, the concentrate was railed to Birkenhead.

By the early 1970s, however, sulphur was being purchased at relatively low cost from Canada, where the refinement of sour (hydrogen sulphide rich) natural gas had resulted in large surplus stockpiles. The sulphur subsidy was abolished in 1966 and production ceased in May 1972. By this time, over one million tonnes of concentrate had been produced from five million tonnes of ore.

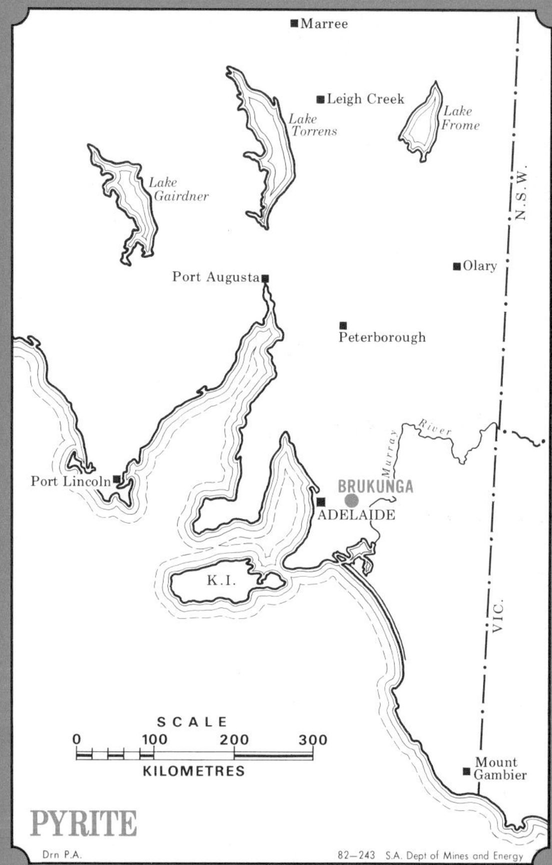

389 Initial stripping of overburden at Nairne Pyrites Ltd Brukunga Quarry in 1952.

390 Diamond drill testing of pyrite ore at Timmins Hill in 1968. The view is towards quarries at Shepherd's and Grassfire Hill, with Mount Barker in the distance.

391 Nairne Pyrites Ltd concentrating plant, tailings dam, and township of Brukunga, circa 1972.

392 Shovel loading a twenty tonne truck with pyrite ore at Brukunga. To the right is a small shovel converted to a dropball by addition of a boom, large steel weight, and rubber tyre shock absorber; this was used to break up oversized boulders prior to trucking.

393 Southwesterly view of Brukunga township and pyrite quarry in February 1982. The quarry is almost two kilometres long and up to eighty metres deep. The remains of the concentrating plant are between the two large overburden dumps, and the tailings dam with its tree screen is at left.

JADE

Nephrite jade was discovered twenty kilometres north of Cowell by Harry Schiller in 1965. Geological mapping has identified more than one hundred lenses of jade over an area of nine square kilometres. Cowell Jade Pty Ltd was formed in 1976 to mine, process and market the product.

Jade is sawn and graded according to colour and quality, and a variety of jewellery, ornaments and carvings are produced at the Cowell and Unley workshops.

To the end of 1981, approximately twelve hundred tonnes of jade had been mined. Dark green jade was considered to be the highest grade at Cowell until black and translucent very pale green were found in 1982.

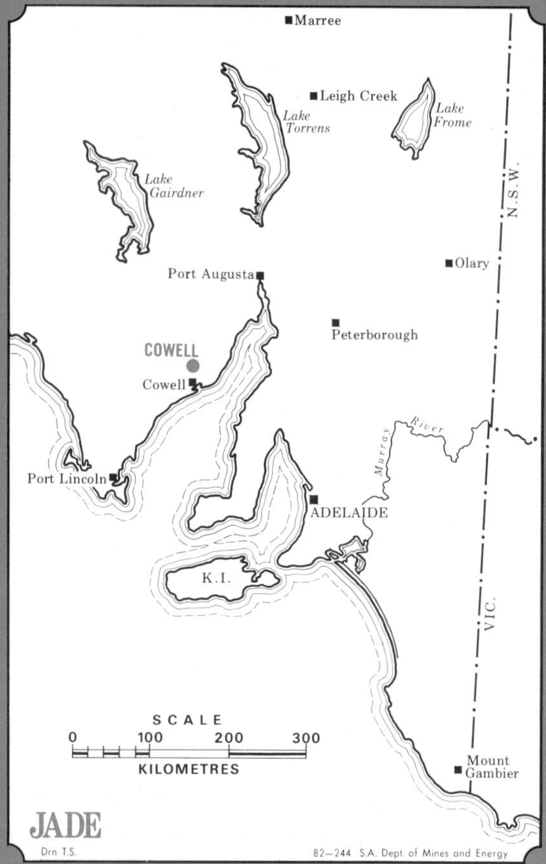

394 A distinctive jagged outcrop of nephrite jade near Cowell. The largest lens found is sixty-five metres long and the smallest a mere twenty centimetres.

395 Hand picking of jade from a quarry face in 1979. The rough surfaces disguise the rock's true beauty, and quality cannot be judged until a face is cut with a diamond saw.

396 Twenty-five centimetre high Gothic harp of black Cowell jade and sterling silver made by Albert Vandyk at the O'Halloran Hill College of Further Education in 1981. The harp took fifteen hundred hours to complete, and is valued in excess of $10 000.

ACKNOWLEDGEMENTS

In research and compilation of this volume, the author has received assistance and encouragement from many sources. Access to photographic and data files, so freely provided, is appreciated.

Many prints have been taken from collections where the name of the photographer has been lost over the years. To these people, the reward for their foresight in recording part of our mining history is in the presentation of this volume.

In particular, the author acknowledges assistance provided by Adelaide Brighton Cement Ltd, The Art Gallery of South Australia, South Australian Archives, Ian Auhl, The Broken Hill Associated Smelters Pty Ltd, The Broken Hill Pty Ltd, Burnside Public Library, Joy Chilman, Philip Driver-Smith, Max Farrell and Associates Pty Ltd, ICI Australia Operations Pty Ltd, Marchant Photographic Services Pty Ltd, The National Trust of South Australia, Robert Noye, Roxby Management Services Pty Ltd, Santos Ltd, and colleagues within the Department of Mines and Energy.

The plans were drawn, and chapter layout designed, by the Publication Drafting Section, Department of Mines and Energy.

PHOTO NUMBER	SOURCE	SADME FILE NUMBER
1	Glen Osmond Institute	33210
2	Sidney Langley	33215
3	Burnside Public Library	33209
4	Sidney Langley	33205
5	S.A. Archives	32123
6	SADME	32795
7	SADME	32176
8	S.A. Archives	32708
9	S.A. Archives	32998
10	SADME	468
11	SADME	1137
12	SADME	33342
13	S.A. Archives	32694
14	SADME	33198
15	SADME	32174
16	SADME	27859
17	SADME	491
18	S.A. Archives	32706
19	The BHAS Pty Ltd	33103
20	The BHAS Pty Ltd	33104
21	The BHAS Pty Ltd	33105
22	SADME	33039
23	SADME	33041
24	S.A. Archives	32704
25	The BHAS Pty Ltd	33109
26	The BHAS Pty Ltd	33106
27	The BHAS Pty Ltd	33107
28	The BHAS Pty Ltd	33102
29	SADME	12902
30	S.A. Archives	32705
31	The BHAS Pty Ltd	33110
32	SADME	24005 24103 24259
33	S.A. Archives	24004
34	SADME	24003
35	SADME	995
36	Collection: The Art Gallery of South Australia	33327
37	S.A. Archives	18857 21382
38	S.A. Archives	18860 18861
39	S.A. Archives	12974
40	S.A. Archives	12973
41	Robert Noye	33245
42	Ian Auhl	32155
43	Adelaide & Wallaroo Fertilizers Ltd	32144
44	Robert Noye	32148
45	Ian Auhl	32156
46	Ian Auhl	32158
47	SADME	12817
48	SADME	32184
49	SADME	32185
50	SADME	32442
51	SADME	32967
52	SADME	33239
53	SADME	33240
54	SADME	22122
55	SADME	33241
56	S.A. Archives	33235
57	SADME	6703
58	SADME	32790
59	SADME	33340
60	S.A. Archives	32725
61	S.A. Archives	32700
62	SADME	11239
63	S.A. Archives	32716
64	SADME	10757
65	SADME	10765
66	SADME	10767
67	SADME	7157
68	S.A. Archives	32714
69	SADME	10762
70	S.A. Archives	32729
71	S.A. Archives	32721
72	SADME	30186
73	SADME	7169
74	SADME	11487
75	SADME	10756
76	SADME	32182
77	SADME	30175
78	S.A. Archives	32722
79	SADME	28583
80	SADME	32181
81	S.A. Archives	32719
82	S.A. Archives	30181
83	S.A. Archives	32737
84	S.A. Archives	32738
85	SADME	7170
86	SADME	7167
87	SADME	11226
88	SADME	10758
89	SADME	20512
90	SADME	32793
91	SADME	10755
92	SADME	20505
93	SADME	490
94	SADME	20500
95	S.A. Archives	32762
96	SADME	33175
97	S.A. Archives	32759
98	SADME	33172
99	S.A. Archives	32763

PHOTO NUMBER	SOURCE	SADME FILE NUMBER
100	SADME	33174
101	S.A. Archives	32760
102	S.A. Archives	32761
103	Philip Driver-Smith	32126
104	SADME	3221
105	S.A. Archives	32768
106	SADME	5234
107	Philip Driver-Smith	32125
108	Jan Tarr	32692
109	Jan Tarr	32693
110	SADME	4584
111	SADME	55
112	SADME	33341
113	S.A. Archives	32960
114	SADME	33412
115	Roxby Management Services Pty Ltd	33363
116	Roxby Management Services Pty Ltd	33362
117	Roxby Management Services Pty Ltd	33364
118	Roxby Management Services Pty Ltd	33437
119	Roxby Management Services Pty Ltd	33438
120	S.A. Archives	32958
121	S.A. Archives	32957
122	S.A. Archives	32948
123	S.A. Archives	32950
124	S.A. Archives	32951
125	S.A. Archives	32959
126	SADME	33312
127	S.A. Archives	32955
128	S.A. Archives	32952
129	S.A. Archives	32956
130	S.A. Archives	32753
131	S.A. Archives	32752
132	SADME	24062
133	SADME	4734
134	SADME	33343
135	SADME	5528
136	SADME	4391
137	SADME	4390
138	SADME	1688
139	SADME	32188
140	S.A. Archives	32754
141	S.A. Archives	24057
142	S.A. Archives	24059
143	S.A. Archives	24065
144	S.A. Archives	24061
145	S.A. Archives	24060
146	S.A. Archives	24058
147	S.A. Archives	32755
148	S.A. Archives	32756
149	SADME	6681
150	SADME	716
151	SADME	26494
152	SADME	26487
153	SADME	26495
154	SADME	425
155	SADME	39
156	SADME	560
157	SADME	1680
158	Ian Auhl	32172
159	Ian Auhl	32173
160	S.A. Archives	32766
161	SADME	32194
		32195
162	S.A. Archives	32229
163	Quarry Industries Ltd	33223
164	Quarry Industries Ltd	33225
165	Quarry Industries Ltd	33224
166	SADME	33248
167	SADME	28392
168	SADME	31033
169	SADME	30234
170	Quarry Industries Ltd	33246
171	SADME	33226
172	SADME	33236
173	SADME	33405
174	SADME	33406
175	SADME	185
176	SADME	471
		472
177	SADME	364
178	SADME	6995
179	S.A. State Promotion Unit	32788
180	S.A. State Promotion Unit	32789
181	SADME	33336
182	SADME	33337
183	SADME	1019
184	SADME	991
185	SADME	989
186	SADME	9402
187	SADME	6920
188	SADME	3746
189	SADME	482
190	SADME	26652
191	Martin Dunstan	32335
192	Martin Dunstan	32330
193	Martin Dunstan	32334
194	S.A. Archives	32702
195	SADME	484
196	SADME	114
197	SADME	32974
198	SADME	32975
199	Adelaide Brighton Cement Ltd	33358
200	Adelaide Brighton Cement Ltd	33359
201	SADME	5070
202	SADME	5290
203	SADME	5069
204	Adelaide Brighton Cement Ltd	33402
205	ICI Australia Operations Pty Ltd	33100
206	SADME	2748
207	Adelaide Brighton Cement Ltd	33361
208	ICI Australia Operations Pty Ltd	33098
209	SADME	32973
210	SADME	729
211	SADME	732
212	SADME	5855
213	SADME	2504
214	Adelaide Brighton Cement Ltd	32691
215	S.A. Archives	32701
216	SADME	144
217	SADME	8578
218	The BHP Co. Ltd	33016
219	SADME	11800
220	SADME	33404
221	The BHP Co. Ltd	33017
222	SADME	412
223	SADME	1640
224	SADME	164
225	SADME	4866
226	SADME	11482
227	SADME	5559
228	SADME	5568
229	SADME	5563
230	SADME	3839
231	S.A. Archives	32776
232	SADME	32971
233	The National Trust of SA	33182
234	The National Trust of SA	33183
235	The National Trust of SA	33181
236	SADME	33184
237	SADME	988
238	SADME	2426
239	S.A. Archives	32961
240	ICI Australia Operations Pty Ltd	33093
241	SADME	33237
242	ICI Australia Operations Pty Ltd	33179
243	SADME	32174
244	Jean and Douglas Giles	33055
245	The BHP Co. Ltd	33014
246	SADME	28775
247	SADME	307
248	S.A. Archives	32745
249	SADME	1671
250	SADME	11436
251	SADME	13046
252	Santos Ltd	33299
253	Jean and Douglas Giles	33054
254	SADME	33391
255	SADME	33389
256	SADME	33390
257	SADME	33422
258	The BHP Co. Ltd	18286
259	Advertiser Newspapers Ltd	33304
260	SADME	14662
261	SADME	83
262	SADME	4268
263	SADME	16588
264	SADME	13483
265	SADME	33418
266	SADME	33408
267	SADME	33416
268	ETSA	32744

PHOTO NUMBER	SOURCE	SADME FILE NUMBER
269	ETSA	32743
270	SADME	33419
271	SADME	33238
272	SADME	2988
273	ETSA	21623
274	SADME	21317
275	SADME	33310
276	SADME	365
277	SADME	329
278	SADME	369
279	ETSA	32786
280	SADME	32373
281	SADME	32980
282	SADME	32979
283	SADME	8104
284	SADME	5281
285	SADME	5179
286	SADME	5178
287	The BHP Co. Ltd	33371
288	The BHP Co. Ltd	3532
289	The BHP Co. Ltd	33019
290	The BHP Co. Ltd	16555
291	The BHP Co. Ltd	33021
292	The BHP Co. Ltd	33013
293	The BHP Co. Ltd	33008
294	The BHP Co. Ltd	33024
295	The BHP Co. Ltd	33023
296	The BHP Co. Ltd	18283
297	The BHP Co. Ltd	33012
298	SADME	62
299	SADME	64
300	SADME	66
301	SADME	11805
302	SADME	31890
303	SADME	1556
304	SADME	5355
305	SADME	2208
306	SADME	2516
307	SADME	2511
308	SADME	13381
309	SADME	17776
310	SADME	10897
311	SADME	33213
312	SADME	33214
313	SADME	715
314	SADME	4864
315	SADME	332
316	S.A. Archives	32779
317	SADME	4276
318	SADME	3334
319	SADME	1564
320	Waratah Gypsum Pty Ltd	33439
321	Waratah Gypsum Pty Ltd	33440
322	Waratah Gypsum Pty Ltd	33441
323	Waratah Gypsum Pty Ltd	33442
324	SADME	11784
325	SADME	33367

PHOTO NUMBER	SOURCE	SADME FILE NUMBER
326	SADME	33409
327	SADME	33410
328	SADME	33414
329	SADME	33411
330	SADME	32175
331	SADME	71
332	S.A. Archives	32709
333	S.A. Archives	32781
334	SADME	1150
335	S.A. Archives	32780
336	SADME	998
337	SADME	33189
338	SADME	33188
339	SADME	9968
340	SADME	22554
341	SADME	32977
342	SADME	4742
343	SADME	33169
344	SADME	9788
345	SADME	6706
346	SADME	5579
347	SADME	4395
348	Nell Lively	32764
349	SADME	7627
350	SADME	238
351	SADME	3281
352	SADME	31338
353	SADME	30916
354	SADME	30920
355	SADME	30914
356	SADME	30921
357	SADME	33216
358	SADME	30915
359	SADME	32982
360	SADME	2962
361	SADME	2959
362	SADME	19909
363	SADME	32965
364	SADME	30239
365	SADME	31365
366	SADME	31359
367	SADME	1346
368	SADME	7836
369	SADME	8891
370	SADME	5912
371	SADME	5894
372	SADME	7843
373	SADME	7463
374	SADME	7446
375	SADME	29861
376	SADME	32793
377	Philip Driver-Smith	32134
378	Philip Driver-Smith	32128
379	Philip Driver-Smith	32135
380	SADME	1559
381	SADME	2086
382	SADME	9277

PHOTO NUMBER	SOURCE	SADME FILE NUMBER
383	SADME	2721
384	SADME	2884
385	SADME	4111
386	SADME	2782
387	SADME	20899
388	SADME	20575
389	SADME	1694
390	SADME	17779
391	SADME	11166
392	SADME	11162
393	SADME	33242
394	SADME	32190
395	SADME	32976
396	SADME	33114

D. J. WOOLMAN, Government Printer, South Australia